MATERIAL MYSTERY

Material Mystery

THE FLESH OF THE WORLD IN THREE MYTHIC BODIES

Karmen MacKendrick

FORDHAM UNIVERSITY PRESS   NEW YORK   2021

Copyright © 2021 Fordham University Press

All rights reserved. No part of this publication may be reproduced, stored in a retrieval system, or transmitted in any form or by any means—electronic, mechanical, photocopy, recording, or any other—except for brief quotations in printed reviews, without the prior permission of the publisher.

Fordham University Press has no responsibility for the persistence or accuracy of URLs for external or third-party Internet websites referred to in this publication and does not guarantee that any content on such websites is, or will remain, accurate or appropriate.

Fordham University Press also publishes its books in a variety of electronic formats. Some content that appears in print may not be available in electronic books.

Visit us online at www.fordhampress.com.

Library of Congress Cataloging-in-Publication Data

Names: MacKendrick, Karmen, 1962- author.
Title: Material mystery : the flesh of the world in three mythic bodies / Karmen MacKendrick.
Description: First edition. | New York : Fordham University Press, 2021. | Includes bibliographical references and index.
Identifiers: LCCN 2021016624 | ISBN 9780823294541 (hardback) | ISBN 9780823294558 (paperback) | ISBN 9780823294565 (epub)
Subjects: LCSH: Anthropomorphism. | God—Corporeality. | Incarnation. | Resurrection. | Abrahamic religions.
Classification: LCC BL215 .M33 2021 | DDC 211—dc23
LC record available at https://lccn.loc.gov/2021016624

Printed in the United States of America

23 22 21   5 4 3 2 1

First edition

For Jaelynn and Olivia
    With hope that the world will share its delight with you

Contents

Introduction: New Materialism, Old Wisdom   1

1 Complex Truth: Myth, Facts, and Matter   15

2 Adam's Skin: The Strangely Bounded Primal Person   38

3 Limitless Bounding: The Valentinian Body of Christ   59

4 Glorious Return: Resurrected Bodies   92

Afterword   125

ACKNOWLEDGMENTS   133

NOTES   135

BIBLIOGRAPHY   181

INDEX   199

The hardest thing of all is to see what is really there.
—J. A. BAKER, *THE PEREGRINE*

Introduction
New Materialism, Old Wisdom

Material Counterstories

> Eternity: for all its invisibility, we gaze at it.
> Muriel Barbery, *The Elegance of the Hedgehog*

Matter is mysterious. We aren't even sure how to describe it—the stuff that takes up space, brute thereness, the something that is other than nothing, the paired opposite of antimatter?[1] By now the claim that matter exceeds our knowing is hardly new, but it is older than its more recent proponents sometimes acknowledge. The self-conscious material turn in the humanities and social sciences has developed along a wide range of theoretical routes, leading to a number of different, imperfectly compatible theories. Still, those theories do tend to have a few things in common. And those things are widely understood as new: in fact, as explicit rejections of older ideas about matter. Academic "turns" often proceed by this sort of rejection. Not infrequently, though, the turns' senses of history are too short. Some of the same texts and traditions in which we find a religious sense of mystery are also the unexpected sources of the complicated, not at all dismissive materialities that will take up the rest of this book.

Tamsin Jones summarizes the turn to materiality in "the study of religion," one of the places we might least expect to find it. She notes a series of rejections and embraces there, "of the interiority, ideality, and emphasis on transcendence that long held sway in considerations of religion, in favor of exteriority, materiality, and immanence."[2] Both the concern with immanence and the desire to reject a single overarching authority may lead materialists away from a sort of naïve theology, where a vast puppeteering God works from on high to make and manipulate the inert material of the world. As theologian Catherine Keller asks, "Is theological materialism, at least of Christian ilk, an oxymoron?"[3] Under the puppeteer, the answer would have to be yes. Immanence and dispersed authority seem to require atheism. But even within the Abrahamic traditions on which the discipline has focused, the long history of theology includes some very different ideas about divinity. The divine may be altogether immanent to the world. Or it may make the clear binary of immanence and transcendence useless, except perhaps as analytic terms for describing its different aspects. In the texts taken up in the following chapters, divine creativity is never sharply marked off from the world.

Rejection of mystery and matter can go both ways. Proponents of each have, at times, declared the other to be thinking and speaking nonsense, on the basis of there being nothing there—or perhaps no there—to know. Some mystical and meditative traditions have called matter, at least as we generally perceive it, an illusion (as have some nonmystical thinkers; the eighteenth-century Irish bishop George Berkeley came to pure idealism through equally pure empiricism, confusing nearly everybody).[4] Modern and contemporary materialists sometimes respond by regarding mystery as a sort of imaginative fantasy, the superstitious byproduct of inadequate science. In no realm, though, has thought progressed neatly from the simplistic to the sophisticated. Like most such moves, the turn to materiality is a return as well as an emergence, and unexpected insights can emerge from returning. The interconnected histories of theology, science, myth, and philosophy can provide us with renewed ways of thinking about our material world and about the astonishment to be found in it, where matter and divinity do not simply come apart and sort out into different disciplines.[5] Drawing upon the rich history of other options will enrich our present, too.

It may be important to many contemporary materialists to reject theism. But for most, it is more urgent to reject a set of anthropocentric biases that have read the world as made for humans, understood by humans, and existing for human use.[6] Instead of a human-centered world, materialist understandings propose an altogether decentered set of relations, whether of energies or systems or objects, with humans no more special than any other point or thing within the set: "Rather, what we get is a redrawing of distinctions and a decentering of the human," writes Levi Bryant, an object-oriented ontologist.[7] New materialist Diana Coole similarly joins philosopher Giorgio Agamben and biosemiotician Jacob von Uexküll to celebrate the "unreserved abandonment of every anthropocentric perspective."[8] Contemporary materialisms challenge humans to take themselves out of the position of Most Important.

For a religious contrast, consider the first creation story in Genesis. Humanity appears there as the pinnacle of creation, made in the divine image, put into the world only after that world has been made good for human habitation.[9] The second creation story there offers a different but still human privilege, in which a part of the desert world is transformed into a lush garden in order to meet human needs.[10] It may thus seem a bit perverse that I will begin my considerations of matter with a set of human-like bodies directly related to these stories—bodies that themselves center myths of creation, redemption, and resurrection. In each case, though, the story that begins as a human story about an anthropomorphic body, though it remains a story told and heard by humans, turns out not to be neatly centered at all. The human uniqueness in the cosmos never holds. Nor could it, because in these readings divinity reaches throughout the world. Like so many stories, the creation narratives in Genesis are stranger and richer than they first seem.

Among the modes of human-centeredness that materialisms take on, those of knowing and doing are especially objectionable. The very fact that we can call a group of disciplines "humanities" tells us that they approach all knowledge from the human perspective. This is not unreasonable; we can try to take the perspectives of other things, but the effort is guesswork at best, and at worst becomes a sort of arrogant colonizing. (This is not to deny that even an attempt at empathy over such a distance can valuably unsettle our selfishness.) More problematically, across all disciplines we have tended to assume that knowledge itself is particularly (often uniquely) human. We've assumed that humans' knowing the world is not reciprocated by the world's knowing humans. And we've assumed that our consciousness—a crucial aspect of the interiority that Jones names as a point of rejection—makes us special. Again, there is considerable variety in the ways that materialisms reject this notion, but they have that rejection in common. The strongest version probably comes from object-oriented ontology, which insists that there are real objects (more of them than we usually imagine), but that these objects' reality is inaccessible to humans—"the things-in-themselves remain forever beyond our grasp," as Graham Harman writes.[11] Harman even insists that our senses fail to give us any access to what is real ("nothing sensual is ever real, no matter how many effects it might have"), and that "all images are false."[12] This is an epistemological claim curiously close to the widely received version of Platonism, where "Platonism" is read as an ancient affirmation of pure immateriality and a rejection of what is sensually perceived. Object-oriented ontological theorists often use the Heideggerian language of objects "withdrawing" from our knowing. In fact, Timothy Morton writes, for any object "we can never see the whole of it, and nothing else can either."[13] This last clause removes the option of a God's-eye view that might supplement limited human perception. Though other modes of materialism are less mistrustful of the senses, they are no less committed to taking away the special privilege of a conscious human surveying the inert material world. Coole and Samantha Frost argue that new materialism, which sees the world in constant, fluid interchange, is in part "a *methodological* or *epistemological* reorientation" away from human-centeredness.[14]

Once more, Genesis seems to offer a strong contrast. There the superiority of human intellect is shown by Adam's ability to name all the animals, not one of whom is allowed to name him in turn. Genesis suggests that even God eagerly awaits the surprise of these names' revelation, while the similar story in the Qur'an has God tell the human alone, not even the angels, what the names are, so that the angels themselves will have to be in awe of the man God has made.[15] As Patricia Cox Miller has pointed out, Christian church fathers made much of the Genesis passage on naming in their enthusiastic declarations of the superiority of the human intellect, which they saw in turn as a clear sign of our place at the pinnacle of creation.[16] The same traditions, though, emphasize the necessary humility of human knowing in comparison with the divine. The very God who invites that naming also warns humans against trying to know too much.[17] Their punishment for doing so anyway, expulsion from their garden paradise, becomes the paradigm of all human suffering.

It is not clear what kind or object of knowledge might be too much. But as Christianity in particular develops, it brings the traditions of the Hebrew Bible together with those of Greek philosophy, sometimes following the work of Jewish Platonists such as Philo of Alexandria. Though Plato is very fond of knowledge, he and his followers always insist that human knowing is narrowly limited in comparison to all possible (and especially divine) knowing. Thus the instruction against knowing too much meets the necessary impossibility of doing so anyway. The two come together to warn against arrogance. Plato's insistence on this caution begins in an early dialogue, the *Apology*, where Socrates famously declares that his own wisdom is the knowledge of his limits and that other so-called wise people lack this humility. In the Academy after Plato, Academic skepticism embraces dialogue, Plato's favored form, as the means of preserving indeterminacy—it can be hard to see who "wins" Plato's arguments precisely because such winning is not the point. Shortly past the turn of the common era, neo-Platonic theurgy sees the highest knowledge as opening onto mystery, so that the most important kind of knowing is the ability to recognize the unknowable in what we know.[18]

The rest of this book considers this unknown—this mystery, secrecy, or simply limitation. Such an emphasis keeps the decentered human from at least some kinds of presumption about knowing. To be sure, there is a real difference between saying that human knowledge is limited in comparison to divine knowledge, which is Socrates' point, and saying that human knowledge does not have special, dominant access in comparison to anything else in our world. The two may come together, though, in theologies of immanent divinity. If divinity is within the world, and the divine eludes full human knowing, so too does the world. The following chapters will look at texts and traditions that value the peculiar knowledge that much is not just unknown, but unknowable, incomprehensible. Such traditions may go on to disperse that mystery throughout the world, resisting the dominance of human knowing over the rest of materiality. Humans can be separated from materiality neither as knowers nor as things that are known. Matter remains mysterious.

The presumption of unique knowledge is caught up with that of unique agency. That is, we have a habit of assuming that humans are subjects who can act in a world full of objects that can be acted on. The matter making up those objects is correspondingly presumed to be inert and passive. Again, Jones is succinct about new materialism, which "emerges, in part, out of the awareness that matter is not in fact dormant, passive, inert stuff that gets acted upon."[19] Jane Bennett describes her own contemporary materialism in similar terms:

> Matter is not the brute stuff for our vital projects or the inorganic obverse of life but is itself a lively actor in the world. This is a materiality whose vitality infects and infuses our bodies, our tools, and our institutions even as it also disrupts, vies against, exceeds, and sometimes destroys them. This materiality is not inert, brute, raw, or capable only of the motion of bare repetition or mechanical reproduction.[20]

Instead, it displays "the freedom, the energetic fluidity or surprising creativity of the natural world."[21] Matter is "lively," as Frost and Coole put it; it "exhibit[s] agency."[22] Such a conception "disturbs the conventional sense that agents are exclusively humans, who possess the cognitive abilities, intentionality, and freedom to make autonomous decisions and the corollary presumption that humans have the right or ability to master nature."[23] Karen Barad similarly observes, "One cannot take for granted that all the actors, actions, and effects are human."[24]

Arguing against the same claim for unique agency, but from a different angle, some speculative realists (a broad heading usually inclusive of object-oriented ontology) claim that to call something real is simply to say that it affects other things—that is, everything real is an agent.[25] Explaining this stance, Steven Shaviro writes, "It is only an anthropocentric prejudice to assume that things cannot be lively and active and mindful on their own, without us."[26] To be sure, some versions of this causality do become very odd. As Morton explains it, "Objects encounter each other as operationally closed systems that can only (mis)translate one another. . . . Causation is thus vicarious in some sense, never direct."[27] In this strange closed-off indirection, humans are just objects among others. Actor-network theory, an intermixing of the natural and social perspectives that is a predecessor to some forms of materialism and realism, gives no more weight to human than to any other agents in any set of relations.[28]

We can stay with the opening stories in Genesis to find apparent counterclaims. Perhaps because of being uniquely in the divine image, humans in the first creation story are granted dominion over the rest of the world, a scriptural moment with famously disastrous consequences.[29] In the second account, once placed in the garden, humans are instructed to "till and keep" the land, a gentler but still dominant relation.[30] There are things in the stories that affect humans. There are trees, for instance, whose fruit can grant knowledge or even immortality. But it is still human agency—deciding, reaching, eating—that enables those effects, rendering the things not agents themselves, but tools for gods and humans to use.[31] There are other agents at work, though: mediating between the humans and the trees, there is a very agential serpent. And while the snake and the humans alike are later condemned to return to the earth, that earth, as we shall see in Chapter 2, itself has quite a lot of agency in their creation. Indeed, a world full of divinity, whatever that turns out to mean, is unlikely to be inert and passive. Though none of the explorations here involves a deterministic theology, human agency is seldom particularly strong—not because it is overwhelmed by a still more dominant god, but because acting is distributed through the world.

It is clear by now that the naïve reading of Genesis that I have put forth is something of a caricature, but it is a recognizable one. The same is true for the received image of Platonism, which we can now recognize as greatly oversimplified in its insistence that Plato is hostile to material things. Genesis seems to place humankind at the center of the world, in the position that God most values, but it opens onto far more interesting options. Plato seems to argue that the immaterial soul is distinct from and superior to the fleeting flesh, a reading that some Christianity will unfortunately tack onto its biblical anthropocentrism. It is thus all the more delightful to see how intensely some quite Platonized modes of late ancient and medieval religious thought trouble the central and arrogant role of humankind, value the world as matter, and make immanent not just the value, but the divinity itself, of that world.

More specifically, the marvels and strangeness immanent to the material world are most readily retrieved from texts that are sometimes within, and sometimes productively tangential to, the traditions of creative Wisdom, especially in Jewish and Christian thought. These are not inherently Platonic traditions—many of the Hebrew Wisdom texts precede Plato. But in subsequent philosophy and theology, Wisdom comes to be entangled with Platonic ideas of goodness and beauty as creative forces that permeate the world. And these mingled traditions offer us a corollary to the mysteriousness of matter: mystery, they suggest, is material.

Wisdom texts are widespread in the ancient Near East, where they frequently offer advice for living well. They may date as far back as the *Epic of Gilgamesh*, in which both secret divine and profound human wisdom are offered to the protagonist—and where wisdom, as Karel van der Toorn points out, might ultimately mean the appreciation and enjoyment of this world's beauties and pleasures.[32] Works of Wisdom appear in Egyptian as well as Mesopotamian literature and in the ascetic Jewish texts from Qumran.[33] The Hebrew Bible contains several books that are categorized as Wisdom literature, though the categorization is inexact and debated. Two of them—the Song of Songs and Ecclesiastes—chiefly offer advice.[34] Job also gives advice. As Dianne Bergant points out, though, it does so through a set of questions about creation and the strange experience of trying to answer them. She offers an intriguingly new materialist possibility: "This breathtaking, even mystical, experience of creation has catapulted [Job] out of his narrow confines of anthropocentrism into the vast expanses of mystery. It has brought him to realize that human history unfolds within the broader context of the natural world, and not vice versa. Job comes to see that the natural world does not merely serve the ends of human history."[35]

The other canonical Wisdom books dwell more obviously on Wisdom's creative role—a role by which wisdom is continuously offered to those in the world. Creation, that is, itself offers the advice we need; the world itself is made such that it is wise and can pass along its wisdom. But we are often inattentive to it or illiterate in the ways it could be understood. Among these books, Sirach is an exegesis of Genesis in terms of Wisdom, probably from the early second century BCE, while the book of Wisdom and some of the Psalms and the Proverbs are probably written closer to the composition dates of Genesis. A lovely passage from the last of these describes Wisdom's role in creation:

> Ages ago I was set up, at the first, before the beginning of the earth. When there were no depths I was brought forth, when there were no springs abounding with water. Before the mountains had been shaped, before the hills, I was brought forth—when he had not yet made earth and fields, or the world's first bits of soil. When he established the

heavens, I was there, when he drew a circle on the face of the deep, when he made firm the skies above, when he established the fountains of the deep, when he assigned to the sea its limit, so that the waters might not transgress his command, when he marked out the foundations of the earth, then I was beside him, like a master worker, and I was daily his delight, rejoicing before him always, rejoicing in his inhabited world and delighting the human race.[36]

Far from transcending the world, Wisdom suffuses it with divinity. She benefits those who will inhabit the world, but she also works to delight her co-creator in the way that she forms the earth and seas. She is the co-creator's delight *by* rejoicing before him—that is, her delight is contagious. It expands from her to those in the "inhabited world," and the world itself is that in which she rejoices. The world becomes lively, conveying divinity by delight. The Proverbial passage sounds at first like a neat division of creator, Wisdom, world, and knowing inhabitants, but the more carefully we think about it, the less tidy these divisions seem. A world infused with Wisdom cannot fail to act on those in it, and if divinity is throughout the world, then any claim to human domination and uniqueness is undercut. In Christianity, a Wisdom-like role often goes to Christ, as co-creative instructor.[37] The canonical Gospel of John, not accidentally like Genesis, begins with the world's beginning, which happens through the *Logos*—the first and formative principle and the speech of God all at once, widely identified with Christ. "In the beginning was the Word [*Logos*], and the Word was with God, and the Word was God," the gospel begins. "He was in the beginning with God. All things were made through him. . . ."[38] Like Wisdom, this co-creative abstraction moves into the world, as we are told later in the same chapter: "And the Word became flesh and lived among us."[39]

As we shall see in Chapter 1, the idea that matter and divinity are closely connected, in ways both straightforward and mysterious, is at least as old as Western philosophy. Especially in late antiquity, the connection becomes vivid in monotheistic religious thought as well. Virginia Burrus writes of what she calls the "ecopoetics" of ancient Christian thought, and she notes several sets of false distinctions by which we misconceive such thinking. "The boundaries between Christianity and what we ourselves might think of as its religious 'others' were not always as evident to denizens of the ancient Mediterranean as they are to us," she points out, noting that the era's "Christian" thought includes aspects that are "also Jewish and Platonic and polytheistic," and all the more "Christian" because of it. Ancient Mediterranean Christianities, she adds, left "the definition of what is properly Christian open"; they "did not oppose science to belief, reason to faith" but created cosmologies that "were creative exercises of intellect and imagination, at once scriptural and philosophical, pushing the boundaries of what thought could think." Importantly for their contemporary relevance, she adds, "Ancient Mediterranean Christianities experienced humans as coexisting with a wide range of other lively, relational beings, whether animal, vegetable, or mineral, angelic or demonic, divine or creaturely, large or small. This ancient ecopoetics is both materialist and animistic."[40] Catherine Michael Chin writes that in late ancient thought, "the cosmos both contained rationality and in some accounts was itself a rational organism. The presence of multiple knowing agents in correspondence made knowledge of the cosmos a matter not merely of human knowledge about the cosmos but of the cosmos itself knowing."[41] Human knowledge is simply part of cosmic knowing. Such thinking, with its disrespect for boundaries, is not a muddled failure to see where the roles and limits of knowers and actors belong. It is, instead, both complicated and complicating.

As Burrus's remarks indicate, late antiquity was a particularly exuberant intellectual moment in Mediterranean thought, if we may be allowed moments that are centuries long. A similarly vibrant period occurs in the Middle Ages, especially as Aristotelian thought is reintroduced to Europe through its Islamic interpreters. Though religious distinctions and boundaries are more marked here than they would have been in antiquity, the intellectual interchanges are no less exciting. Both are moments rich with creatively heterodox theologizing. They are moments in which exegesis is especially prevalent and especially lively, and new myths are created as ways of reading older ones. As we might therefore expect, they are not moments with a particular concern for conceptual purity. They involve not only mutual influence among religious traditions, but also several kinds of Greek philosophy, and an occasional trace of older Mediterranean philosophies and religions too.[42] There is a great deal of scholarly work on the evidence for these influences, and speculation about its direction—that is, about who got which ideas from whom. It would be absurd to deny the interest and importance of this work, but I will not be drawing on it very much, engaging instead with a few of the ideas and images that roam across that range of traditions and are particularly relevant to considerations of matter.

In such intellectually exuberant moments, traditions and ways of reading interpenetrate. Like the widely varied versions of contemporary materialism, readings influenced by ideas of Wisdom share some common concerns, grounded in a sense of the world's remarkable richness. Perhaps most importantly, they focus on creation (both the creating process and the created world) and on the instructive power that is part of the activity of the world itself.[43] It is *because* of its role in creation that Wisdom has the role of instructor.[44] Wisdom is throughout the world; the world is wise throughout. Its matter has things to teach us, as part of it.

In keeping with the indistinction of the boundaries of Wisdom literature, I will draw here on texts that are interested in material creation and in the figures of created, redeemed, and risen bodies, but are not (always) considered to be Wisdom texts. Among these, Rabbinic commentaries from the early centuries of the common era may provide an exoteric wisdom that complements the more esoteric wisdom of groups like those who wrote the texts found at Qumran.[45] In these commentaries, Wisdom is not only a scriptural figure, but is at times identical with Torah, the first five biblical books that preoccupied early Jewish commentators.[46] The Torah, like Wisdom, is the way that divine teaching can reach human beings. Some scholars even suggest that the commentaries may be works of Wisdom literature themselves.[47] Ishay Rosen-Zvi points out that the rabbis, who were "notoriously reluctant to adopt other biblical genres," nonetheless produced "wisdom-literature styled aphorisms." He hypothesizes that perhaps "the rabbis do so because they view themselves as part of the wisdom tradition."[48]

The rabbinic texts, along with proto-orthodox Christian works, are likely to be the most familiar among my sources. That familiarity is intentional. The writers of both are especially active in the first several centuries of the common era, where the stability of their communities cannot be taken for granted. Those writers work to render their faiths, and correspondingly their communities, as coherent as possible, and they are successful enough to establish the nature of mainstream Christianity and Judaism after them. Both work from canonical scripture, much of which is shared—though the rabbis are also engaged in extensive interaction with oral traditions of interpretation, and Platonic, Aristotelian, and Stoic thought have already entered Christian thinking. The rabbis are less concerned than the Christians are with establishing the bounds of their communities by exclusion, though they do condemn wrong opinion in their arguments. These sources are familiar, then, both because they came to be dominant in their respective faiths and because they intentionally sought to become familiar, to reach and to convince as many as possible.

The early Christian interest in exclusion, as part of the effort to establish orthodoxy, is directed with particular intensity at the groups the heresiologists called gnostic. The term is based on those groups' focus on salvation through *gnosis*, a kind of knowing that generally has an element of mystery. There may be one group, much like those later called Sethians, who applied the label to themselves, but it is more often applied from the outside, and its boundaries remain contentious and unclear.[49]

The Sethians might have had pre- or non-Christian versions, but they adapted their mythology to a recognizably Christian form, though the members of their Trinity were Father, Mother, and Child, the Mother joining with other feminine Wisdom figures in both creative and salvific roles.[50] The most elaborate and sustained denunciations, though, were directed not at the Sethians but at the Valentinians, who inspired Irenaeus's intense and lengthy polemic *Against Heresies*. Irenaeus in turn inspired most of the heresiologists who came after him. The Valentinians emphatically considered themselves Christians, and, as Ismo Dunderberg notes, "It was difficult for ordinary second-century Christians to differentiate between Valentinians and other kinds of Christians."[51] That, of course, is probably one reason that the heresiologists were so upset by them. It wasn't just "ordinary" Christians who would include Valentinus and his followers among themselves, either; influential Christian writers, such as the Alexandrians Origen and Clement, engaged with them as serious Christian thinkers.[52] On the Valentinian understanding, Wisdom has a complex role, working to create and to save in complicated interaction with different figures of Christ. The Valentinian relation to materiality, and that of the Sethians, too, have long been thought of as hostile ones, but this is hardly the only possible reading.[53]

Elements of elaborate gnostic cosmologies appear again in medieval and early modern Kabbalah, which weaves together extensive knowledge of both rabbinic interpretive argument and intricate Platonic concepts of the cosmos and its creation. Though we do not find a more traditional role of Wisdom in creation and redemption, we do find a divine presence scattered throughout the world. Kabbalistic texts are sometimes read as urging an ascetic detachment from the world of matter, or even suggesting that it is evil. Here again, without delegitimizing those readings, we can find traces of other possibilities—not surprising in such complicated texts.

Many of the texts and traditions that touch upon Wisdom are intentionally mythic and poetic. Others appear more straightforwardly philosophical, or even logically argumentative, but these too tend to make extensive use of metaphor and small stories, weaving poetry into reason. After a consideration of the kind of discourse that exegesis may involve, the following chapters approach matter by each beginning with a religious body: the primal human, the divine teacher, and the risen or glorious body. This clearly risks returning to anthropocentrism, with all its complacencies. But from a closer consideration of this trio of strange, human-like bodies, something wonderfully odd emerges. Those very bodies trouble the possibilities of neat boundaries and clear edges between world and divinity, divinity and humans, humans and the rest of the world. Each almost-human body turns out to affirm neither an anthropocentric world view nor the superiority of a separately spiritual realm.

A bit more detail may be useful. The starting consideration of discourse is necessary because these claims will not make sense unless we consider the relevant exegeses as matters of myth. The bodies of the primal Adam, the salvific Christ, and the glorious resurrected flesh, the bodies that offer us material mystery, are mythical bodies, and so before we explore them it will make sense to think about the value of mythic discourse. In the process, the received story about the hostility between philosophy, theology, and the love of matter already begins to come apart. So, before turning to these bodies, Chapter 1, following this preface, will explore the story that philosophy emerges out of the struggle against theology and poetry, a struggle in which the material world can be understood scientifically. This means looking both at early Greek ideas about matter, which are even weirder than we might expect, and at the loss of wisdom that comes about in our effort to eliminate myth from truth telling. I want to suggest that the myth, poetry, and theology that philosophy has so long been proud of leaving behind may be vital to understanding the matter of the world; at the same time, what understanding means may be other and more multiple than we have come to think.

As we move into the body of the book, the first figure is that of the first human, or the primal human—sometimes identified with Adam of the Genesis story, sometimes a human-like being created prior to that story. Without being identical to any of them, its body variously maps onto the cosmos, the earth, and the body of God. We can imagine that this makes the whole world and the knowing of it human. But the interpretation likewise inverts itself, so that neither god nor human nor cosmos can quite be sorted out from the matter of which it is made.

The next relevant body is that of the Christian incarnation, where divinity takes on the human body of Jesus. This chapter focuses more narrowly on a single tradition—more exactly, on the particularly Valentinian version of that body and the way that it functions redemptively. There the embodied figure identified as the savior works with Wisdom to teach a marvelously self-complicating perspective on matter. Though there is something of this perspective at work in other theologies with similar philosophical influences, it is seldom as clearly laid out as in Valentinian cosmologies.

Finally, the third body to be considered is the resurrected body that is fundamental both to early Christianity and to some of the rabbinic arguments. Like the primal bodies and the incarnate God, resurrected bodies exceed and disorganize their spatial boundaries and get themselves mixed up, not only spatially but temporally, with all other things.

What emerges from considering embodiment across these lines of thought is not the human made universal, but an astonished sense of the flesh of the world. It is a world of constant becoming, "for," as the book of Wisdom declares, "Wisdom is more mobile than any motion; because of her pureness she pervades and penetrates all things."[54] In the world's wise becoming, the element of beginning, of creating, is especially pronounced. To begin to follow that immeasurable movement, and to watch it display the matter of the world as divine, we can turn to philosophy's own beginning, and to the strange consequences of its efforts to distinguish itself from the myth that pervades and penetrates it.

1
Complex Truth
Myths, Facts, and Matter

The Glorious Birth of Pure Reason

> What I want to return to is the simple straightforward rationality of the Pre-Socratics.
> Karl Popper, 1958 Presidential Address to the Aristotelian Society

Philosophy is etymologically and at its heart the love of wisdom. Other disciplines often begin as parts of it, spinning off into distinct fields as they become less speculative. Throughout philosophy's quite long history, there has been tension regarding the form of the wisdom it loves; in philosophy's usual origin story, this plays out as a tension between reason and myth. Despite recurrent temptations and the occasional slip, says the story, reason wins out. But perhaps we might follow the skeptics who followed Plato, suspecting that the matter is more complicated, and more dialogical.

The usual story begins once upon a fairly particular time. In the twentieth century, Alfred North Whitehead declared that the entire tradition of European-influenced philosophy "consists of a series of footnotes to Plato."[1] Well before Plato wrote his dialogues, the philosopher Anaximander wrote to his pupil Pythagoras, "Let all of our discourse begin with a reference to Thales."[2] Diogenes Laërtius, an invaluable chronicler of early Greek philosophy (though not a perfectly reliable one), traces a line of philosophical development from Thales all the way through to the New Academy of Plato's successors, a range of more than half a millennium.[3] All of Western philosophy, then, might be read as a series of footnotes that begin with reference to Thales. Through this reading, we can trace the heritage of contemporary philosophy back to the sixth-century outburst of rational thought that Thales inaugurated.

For all its varying approaches and uncertain disciplinary limits, Western philosophy has held with remarkable consistency to this story of its origins: in the beginning was Thales, and he set into motion both philosophy and science, moving them away from poetry and theology. Those older discourses were fundamentally obscure and irrational; leaving them behind made truly intellectual pursuits possible. Like so many origin stories, though, this one covers over some marvelous quirks and oddities, which tell us about a world more deeply mysterious and wonderfully strange than we have usually suspected. To make sense of all this, we need to consider the minor matter of truth.

Truth cannot be taken apart from its telling; knowing and saying are tangled together. The pre-Socratic philosopher Parmenides declares, "The same thing is there for thinking of and for being," even while his own strict immaterial monism cheerfully undermines "thing," "there," and "being" in a way that leaves "thinking" pretty thoroughly baffled.[4] Parmenides seems to have denied the reality of matter altogether, and he influenced, among others, Plato.[5] As I've noted, Plato in turn is widely understood as having regarded the material world as being, if not exactly illusory, at least unimportant and untrustworthy.[6] He is also, no less contestably, credited with codifying the distinction between myth and reason upon which philosophy depends. To work out what kinds of things count as true, then, is also to work out the proper modes of saying them.

And first in that line of work is Thales, a polymath from the Greek city of Miletus, in what is now Turkey. He is the earliest among the thinkers broadly known as the Milesian monists, each of whom believes in a single underlying or original substance that makes up the entire world, though they disagree as to exactly what that substance is. Nothing of Thales's own writing survives, if indeed he wrote. We know that, like the more familiar nonwriting philosopher Socrates, he taught and participated in public affairs. Like Socrates, too, he is so widely cited that we can know something of his thought, and still more of what was generally attributed to him. On the basis of his wisdom, he was honored as one of the Seven Sages of ancient Greece.[7] He was said to be an accomplished astronomer who predicted a solar eclipse in 585 BCE, a mathematician who first inscribed a right triangle in a circle, a political advisor who saved the city of Miletus from dangerous alliances.[8] He was, if the stories can be believed, a shrewd investor as well, one whose meteorological understanding enabled him to predict the abundance of the olive harvest, corner the olive-press market, and earn a fortune in rentals.[9] (According to those stories, he performed this feat to demonstrate that his lack of riches was due to disinterest, not inability.)[10] Diogenes even presents several bits of attributed dialogue suggesting that Thales had considerable dry wit; for instance, "when his mother tried to force him to marry, he replied it was too soon, and when she pressed him again later in life, he replied that it was too late."[11]

None of this, however, gives him his status as first philosopher. That derives from his declaration that there is a single source and substance of all that is, a single origin: water. As Aristotle points out, in the text that firmly establishes Thales in this honored position, Thales might have been inspired by observing the necessity of moisture for the generation of plant and animal life, since "that from which a thing is generated is always its first principle."[12] Aristotle considers the history of philosophy, beginning with Thales, by tracing views of causality.[13] The different understandings of "cause" become key in later distinctions between myth and philosophy. Myths, on this dichotomized view, attribute things and events in the world to the whims of external, supernatural agents, while philosophy joins science in seeking those causes within the world of matter. Aristotle argues that the most prized branches of knowledge are abstract, and not utilitarian; they require some leisure—such as that of the Egyptian priests who first created mathematics.[14] Among kinds of knowledge, "Wisdom is knowledge of certain principles and causes," and is the most generalizable and universal kind of knowing.[15] The most universal kinds of causes, in turn, are not those of individual events but of the world as a whole. They include the matter from which things are made, the forms things take, the forces that move them into those forms, and the aims of their existence.[16] For the Milesian philosophers, the *material* cause of things is especially important: what kind of stuff makes up the world? Following suit from the primacy that Thales grants to water, other Milesians will propose different primary materials—misty air, for example, or even an amorphous *apeiron*—but they are materialists all.[17] Following Aristotle, we can see in Thales the use of inference as well as deduction; that is, he makes observations and reasons from them to generalizations, as well as forming arguments from first principles. Thales's innovation is a matter neither of the use of reason nor of the discovery of origins alone. Rather, it is the ability to *generalize to first causes* that sets Thales apart. Causes, universal rules, and material explanations are so important in part because they are origin stories, tales of beginnings, which are as necessary to science as to myth. To present a clear demarcation of philosophy from what he regards as less rigorous thought, Aristotle credits Thales with making a distinct break from earlier cosmological explanations. Aristotle's story of philosophy's beginning remains surprisingly current.

There were plenty of other origin stories before Thales, of course. Perhaps a century before Thales taught, Hesiod had written his *Theogony*, a story of the world's creation, sung by the muses and detailing the making of the cosmos from its beginning in Chaos and Eros.[18] But Hesiod is not considered a philosopher—he is, rather, a theologian and a poet. Chaos, Eros, and the other gods of his epic are clearly mythical figures. Earlier still, Homer seems to have shared Thales's high opinion of water, calling Ocean "the source of all things," but he is not considered a philosopher either—he is a poet whose claim for the ocean is once again grounded in myth.[19] Plato draws our attention to the commonality while still sorting out the poets from the philosophers. In the *Theaetetus*, Socrates declares, "nothing ever is, but things are always coming to be. About this theory, we can assume the agreement of the whole succession of wise men [that is, philosophers], apart from Parmenides—not only Protagoras, but Heraclitus and Empedocles as well; and we can also assume the agreement of the best poets in each genre—Epicharmus in comedy and Homer in tragedy."[20] (To be sure, Socrates and especially Plato will be much concerned with whether the philosophers and poets are right about this.)

Aristotle too is well aware of the apparent continuity across these figures, and it figures into his considerations of philosophy's origins. "All begin . . . by wondering that things should be as they are," he says, and in this way, "the myth-lover is in a sense a philosopher."[21] He even follows Plato's Socrates to note that Homer seems to agree with Thales regarding the primacy of water.[22] What distinguishes Thales, making his a properly philosophical pursuit of first principles, is his move beyond myth, beyond Chaos and Eros and Oceanus and Tethys, beyond the songs of the muses, to a different kind of quest for "knowledge of the primary causes."[23] This new approach, characterized by remaining entirely within the world and by its derivation from observation and reason, marks the emergence of philosophical clarity and no-nonsense physics out of the mythological murk. In her discussion of pre-Socratic cosmologies, M. R. Wright echoes Aristotle in more modern terms: "The early Presocratics realized that there were laws controlling the natural world, and that these were discoverable, so that the whole cosmic structure became the subject of research (*historia*) and rational analysis. . . . They progressed from myths that provided answers that were external and unpredictable (a lightning flash due to the anger of Zeus, the rainbow as a messenger from Olympus) to explanations that were internal, self-consistent, intelligible in natural terms, and predictable."[24] The other Milesian monists, and then the long series of their philosophical descendants, follow Thales in their efforts to find the world's governing principles within the world itself, and not in the realm of gods; in nature, and not supernaturally. They are not mythmakers, but instead searchers for *logos*, for the principle behind the thing. Lucid argumentation replaces poetic tale.

In order to distinguish philosophers' views of causality from theories that he will not count as philosophical, Aristotle formulates criteria for philosophy on the one hand and "myth or theology" on the other. Philosophy's beloved wisdom seeks abstract universals for knowledge's own sake.[25] The stories of the mythographers, on the other hand, "considered only what was convincing to themselves." Such stories make claims "in a sense significant to themselves," Aristotle acknowledges, but "as regards the actual applications of these causes their statements are beyond our comprehension."[26] Taken literally and rationally, the poets' stories make no sense. Both truth and the telling of it distinguish philosophy from myth, and really, says Aristotle, "it is not worthwhile to consider seriously the subtleties of mythologists."[27] Later Theophrastus, Aristotle's successor at the school he established, will confirm Aristotle's claim that Thales's fundamental concern was the investigation of nature, not speculation about divinity.[28]

The story continues: once Thales introduces the approach driven by logic and evidence, that approach transforms the development of Greek thought. Finally, the interfering gods can be removed from the world, no longer needed as means of explanation. Matter can become the subject of physics. So too divine inspiration can be removed from the quest for knowledge, for which nature itself provides more than sufficient motive.[29] Instead of hints breathed into human ears by the elusive Muses, we come to know through the relatively slow but far more reliable processes of reason and observation, through critical inquiry and careful analysis.[30] Instead of discovering Poseidon in conflict with Aeolus and carefully calculating our prayers and sacrifices, we can attend to windstorms over the water and steer our ships accordingly. Patricia Curd explains that the Presocratics distinguish themselves as philosophers by "seeing the world as a *kosmos*, an ordered natural arrangement that is inherently intelligible and not subject to supra-natural intervention."[31] In an etymological hint of poetry's persistence, though, the world's order is also connected to the making of beauty, giving us *cosmetic* alongside *cosmic*, delight alongside intelligibility.

Nature and the gods might be disidentified and removed from mutual influence, but for quite a long time, they continued to coexist. Nonetheless, in *Battling the Gods: Atheism in the Ancient World*, Tim Whitmarsh argues that the new philosophical approach was what eventually made godlessness possible. There were only a few early philosophers willing to follow "the idea that the universe was made out of matter" into "the conception of a god-free reality." But pre-Socratic philosophy, Whitmarsh writes, was important in breaking with the past, "celebrat[ing] the critical spirit, the willingness to question received values." As he points out, "The idea that progress is made by breaking with the past, by rejecting and questioning, is not a self-evident one, and it calls for some explanation."[32] Myth-making and myth-receiving are presumably uncritical and unquestioning; they accept received wisdom with an unprogressive passivity, opposed to philosophy's critical, secular spirit.

It seems odd to attribute passivity to an activity with "making" in it, but while individuals do certainly make or remake myths, myths' creation is often so dispersed across a culture's history that we can readily understand them as a kind of received wisdom. Richard Rojcewicz, in his own consideration of Thales's materialism, puts the distinction neatly: "Myths are handed down; they are traditional, stemming from a remote past. They have no particular author . . . , [while] philosophers are autonomous, their own thinking is the source of their pronouncements. Philosophers rely on nothing but their own eyes and their own reason. Myths offer supernatural explanations of phenomena, whereas the philosopher rejects the supernatural and abides with the natural."[33] A particularly enthusiastic nineteenth-century text even presents Thales as an intellectual liberator: "The same love of liberty, which led him to place himself in this opposition to the prevalent theology, is also shown in his defending the independence of Ionia against Croesus."[34] And so the story goes, on through Greek philosophy and into the Roman Empire, where some of the Greek materialisms (such as Stoicism and Atomism) emerge with renewed vigor. In the process of turning to matter, philosophy turns away from myth, liberates and emboldens the mind, and paves the way for science to come. Philosophy and matter must be saved together, both wrested away from their occupation by interfering gods. Such is philosophy's origin story. But the very fact that it is a story alerts us to its inexactitude.

Shifting into Science

> They say we have woken
> From a long night of magic,
> . . .
>
> Mathematics sinks its perfect teeth
> Into the flesh of space
> But space is unfeeling.
>
> Rebecca Elson, "The Last Animists"

It is unsurprising that Thales is widely credited not just as the first philosopher, but as the first scientist. Both disciplines concern themselves with fundamental causes and natural explanations.[35] And it is important for philosophy that its origin story can be tied to the lasting respectability of natural science.

The two will continue to connect, though more than once, as philosophy drifts toward other disciplines, it will have to be pulled back toward science again. Theology in particular seems to pose a recurrent temptation.

If Thales presides over philosophy's birth, its most dramatic rebirth comes in early modernity, where matter too is reconceived, or perhaps just taken further away from divinity. Late ancient and medieval philosophy do not exclude scientific inquiry, but their theological emphasis is part of what leads to the latter era, with all its intellectual vibrance, receiving the label of the "Dark Ages," on the assumption that religion is amenable to inquisition rather than to inquiry. In the received story, Christianity dominated Europe, and Christianity is especially hostile to nature. This story must ignore not only the presence of intellectual activity, but its sources in the interactions among religious and philosophical traditions. Christianity could not have been so lively except in interaction with the liveliness of Islamic and Jewish thought, and they all would have been less vibrant without their readings of Plato and Aristotle. In the intellectually grim version of the story, though, the dominant Christian claim is that the natural world of matter is inferior and deceptive—it is to be rejected, and in its mingled heritage from Jewish and Greek thought, Christianity comes to insist on a God who is transcendent, immaterial, and demanding. With the rejection of materiality, of course, the rejection of science easily follows—the truths of the natural world are not eternal truths.

If this is the state of learning of the Middle Ages, then a new liberation is urgently needed, and a new philosophical liberator does indeed emerge. This is Rene Descartes, a seventeenth-century, Jesuit-educated French mathematician and physicist.[36] In this part of the story, Descartes rejects the suffocating intellectual dominance of Catholicism, with its emphasis on unquestioning faith, and brings philosophy into a new and lasting era of logical certainty. Despite his own doubts about the reliability of sensory evidence, Descartes was important among those developing Aristotle's insights, including the necessity of establishing causes as the foundation of scientific explanation.[37] (To be sure, Descartes studied Aristotle in the process of his Jesuit education, heavily oriented toward the work of Thomas Aquinas, but that is not part of the way that the story is usually told.)

In a letter to the theological faculty of the Sorbonne, to whom he dedicates his *Meditations on First Philosophy*, Descartes declares that he wants to demonstrate what have traditionally been regarded as religious truths, such as the existence of a single god and the immortality of the human soul, through the use of reason alone, obviating the need for faith in Christian myth.[38] Descartes couches his arguments as a way to persuade unbelievers of eternal truths—but we have to remember that he is well aware of the church's political power.

After all, the *Meditations* were published less than a decade after Galileo was sentenced to a lifetime of house arrest. And while Descartes might have been sincere about his intent, his arguments for God and the soul are neither very impressive nor new.[39] The far more successful element of his discussion is his effort to shift philosophy, once more, from mythic to logical thinking, from scripture or authoritative citation to deductive proof. Descartes, again, is a scientist and a mathematician. He remains committed to the rigor of the inductive and deductive methods of those fields, and so he is well suited to returning philosophy to its (supposed) original intent. In part under his influence, matter comes to be viewed as entirely passive and mechanistic, wholly subject to the laws of nature and to human knowing.

The "scientific method" proper, emphasizing repeatable experimentation along with observation and reason, is inspired in part by Descartes's careful thinking and considerable scientific observation. This method develops in parallel with the long philosophical moment called the Enlightenment. Here is the summary from the Stanford Encyclopedia of Philosophy, generally a straightforward source without much bias or hyperbole: "The dramatic success of the new science in explaining the natural world promotes philosophy from a handmaiden of theology, constrained by its purposes and methods, to an independent force with the power and authority to challenge the old and construct the new, in the realms both of theory and practice, on the basis of its own principles."[40] Like the Milesian story, this one pairs *real* philosophy with science and not with myth: "The rise of modern science in the sixteenth and seventeenth centuries proceeds through its separation from the presuppositions, doctrines and methodology of theology."[41] The language is important. Theology is an enslaving constraint, philosophy is a freeing force; theology is all given as presupposed, philosophy (like science) is an adventurous exploration of the truth as it newly emerges. Theology constrains philosophy; science liberates it.

In the late eighteenth century, Immanuel Kant summarizes Enlightenment philosophy with the modesty characteristic of the era, calling it a release from the immaturity that humankind imposes on itself, an immaturity he describes as "the inability to use one's understanding without guidance from another." Immaturity's self-inflicted cause, he adds sternly, "lies not in lack of understanding, but in lack of resolve and courage. . . . Laziness and cowardice are the reasons why so great a proportion of men, long after nature has released them from alien guidance, nonetheless gladly remain in lifelong immaturity."[42] As cowards, we are fond of "rules and formulas" and bind ourselves to them—particularly in matters of religion, where those with authority are especially concerned to keep the rest of us bound.[43] In 1944, in the *Dialectic of Enlightenment*, Theodor Adorno and Max Horkheimer summarize this point: "Enlightenment's program was the disenchantment of the world. It wanted to dispel myths, to overthrow fantasy with knowledge. . . . Enlightenment, understood in the widest sense as the advance of thought, has always aimed at liberating human beings from fear and installing them as masters."[44] The way to freedom from fear is knowledge itself. "Humans believe themselves free of fear when there is no longer anything unknown," write Adorno and Horkheimer.[45] That is: when there are no longer any mysteries, when there is nothing left for myth to tell, or when the things that only myths and poems can tell have been decisively devalued. Freedom from fantasy means freedom from fear. It means independence of thought, and true knowing.

As long ago as Epicurus, a philosopher about a century Socrates's junior, freedom from fear was understood as requiring freedom from gods. Epicurus believed in gods, but he found it absurd to think that they would trouble themselves with humans.[46] The Enlightenment disinfects reality from gods and spirits: "The disenchantment of the world means the extirpation of animism," as Adorno and Horkheimer explain.[47] Both the liberating agent and the liberated faculty, human reason is at the heart of Enlightenment. We take animation out of the world, and it ceases to be enchanted, to be known and to make itself known by song and story rather than by demonstration and argument. In rescuing the material world from divine domination, we have also rescued ourselves—and put ourselves in charge of the rest of the world.

Though there are some lingering arguments during the early modern era between those who favor reason in its purest abstraction and those who prefer to ground it in the senses, all Enlightenment thinkers believe resolutely in an orderly and law-governed natural world and in our ability to know that world (sometimes within limits) by our own faculties of reason and observation, the very skills with which Aristotle so long ago credited Thales. That considerable good comes from the desire for learning that is clear, fully sharable, and encyclopedic seems to me unarguable. Other parts of reason's successful campaign against myth, however, may demonstrate the losses of wholly disentangling the two—and the fortunate possibility that the effort, in any case, fails.

Reason in Its Usefulness

> Poetry is opposed to science.
> Samuel Taylor Coleridge, "The Definition of Poetry,"
> from *Shakespeare*

The stature of myth, already low as the value of reason rises, worsens as the grounds of reason's value change (though myth does generate a burst of enthusiasm among the Romantics at the outset of the nineteenth century). In their analysis of the desire to know, Adorno and Horkheimer begin to show their readers the shadows cast by this illumination. These largely arise from a change in what Western thought counts as truth, a change that continues to develop over subsequent centuries and is still developing in our own. As this change gets underway, matter is still understood to lack agency (though the understanding becomes far more complex than Descartes could have imagined). But the human attitude to the matter that is not us becomes a bit different.

We might best see the change by looking back. Aristotle, we recall, felt that the highest wisdom was knowledge for its own sake, not subordinated to any utility. He loved understanding, with its attendant pleasures and open horizons (there is always more to learn); he assumed that all people by nature desire to know.[48] This love holds on through early modernity, but not much past it. In contrast, "Modern knowledge is above all technological"—it must be of use. "It aims to produce," Adorno and Horkheimer write, ". . . [not] the joy of understanding, but method. . . . The 'many things' which, according to [Francis] Bacon, knowledge still held in store are themselves mere instruments."[49] As the modern world develops, the most valued thought is instrumental, and the epitome of the instrument may be the hammer. "Only thought which does violence to itself," according to Adorno and Horkheimer, "is hard enough to shatter myths."[50] This violence doesn't seem to have any joy in it, only a grim hostility against the useless. Material things cease to be important simply because they are inherently fascinating, and become either tools themselves or resources to be developed by and for other tools. Humans master the world by using it.

Even if the Milesian turn to reason was a turn away from myth, it did not *destroy* myth, at least not successfully enough for modern tastes. On the instrumental view, myth crept back into our thinking, like a bacterium after an insufficient course of antibiotics. Philosophers learned from this "mistake," Adorno and Horkheimer suggest, and now, "anything which does not conform to the standard of calculability and utility must be viewed with suspicion."[51] Vigilance against inutility is the only way to avoid reinfection. There is no room for what is not factual, and a fact must be provable, quantifiable, and generalizable—not as the object of an Aristotelian desire to know, but so that it may be put to use. In order that all knowledge may be useful, the world must be disanimated, stripped of Chaos and Eros, of mystery and wandering desire, of its highly inefficient capacity to stop and amaze us.[52] Wisdom—a term that now sounds strange, perhaps more amorphous than "knowledge"—is no longer a mobile and pervasive delight. The Enlightenment rebirth of reason must provide a more definitive overturning of theology than we found in antiquity. What begins as a legitimate and urgent objection to the abusive political power of the European church becomes a loathing of any thought that smacks of the sacred. Knowledge is power, and there can be no room in our world for the competing powers of gods. With the gods removed, humanity can take over: "Man's likeness to God consists in sovereignty over existence, in the lordly gaze, in the command."[53]

The story develops easily enough into the assumption that every kind of knowing is trying to be a science, in the Anglophone sense. Social sciences strive to be considered natural, "hard" science. Developments in the sciences draw interest from humanities scholars, but as Marjorie Garber points out, the converse is seldom true.[54] The arts and humanities might be drawn to the sciences out of a happy Aristotelian desire to know, but they also seek to claim some of the rigor that they are no longer perceived to have. Instrumentalism makes all of this worse: the sciences, too, are damaged by the demand that truth be useful. Actual scientists might still delight in knowing, but funding imperatives push research toward utility. The loss of delight meets the demands of production to ensure that thinking, too, is just work. The poetic is frivolous. The mythic is unacceptable. The mysterious is nonsense. And matter is awaiting our use.

When late twentieth-century philosophy once more wanders off from science and comes under suspicion of being far too influenced by literary theory, part of the work of the materialist turn is to reemphasize the things themselves, ideally to be engaged in terms compatible with emerging concepts in physics. Wandering into science is a more acceptable diversion than wandering into poetry and myth, but contemporary science does not reduce materiality to mechanics or even to passivity. As many scholars have realized, there is promise here, in going beyond the disciplinary boundaries between the humanities and the sciences.[55] I would like to suggest that there is promise, too, in the appreciation of myth. These are not the same promise; rather, they are different paths to the shared thought that the material world is stubbornly irreducible to its use value for human agents. Useless matter may be told in nonfactual truth as well as in scientific fact. There is a kind of truth to myth—to poetry and song and story—that is imperfectly predictable, not especially useful, and not reducible to fact, any more than facts can be reduced to mythical tales.

By now we know that gods do not go gently; in fact, they do not go away at all. In our effort to shatter and discard myth, we develop what Bruno Latour calls "the modern cult of the factish gods."[56] Our contemporary popular understanding of "having faith" in gods or myths comes to mean taking them for existents or facts. Factual, useful truth can have no room for mythical methods and their useless revelations. It is clear that entities from myth remain conceptually active, but as our collective sense of truth comes to be more definitively set forth by science and history, we come to affiliate myth with falsehood.[57] That is: to dismiss myth, utilitarian reason first assumes that myth works like fact. Myths are not very good facts, having neither logic nor induction behind them. Thus we have only facts and a poor attempt at facts, so it would be sensible to rid ourselves of the lesser version. This perspective then extends itself backward, to the idea that myths have always been (attempts at) historical accounts and pseudo-scientific explanations. This tale has no room for poetry.

The unfortunate conflation of truth and fact has been used to devalue matter still further. It ends up disregarding not either one, but both ways of truth telling, mythic and factual. It seems to happen because the appeal of mythic stories is powerful, and the identification of truth with facts is very nearly as strong. The effect of this pair is exacerbated as value becomes entirely instrumental. So "believers" can continue to hold onto myths—but they do so by insisting that the myths are factual. In the process, they have to devalue evidence—rational, empirical, or historical. The need for evidence gives way to the usefulness that a proposition has for a given in-group. Without the need for evidence, we imagine that all opinions are created equal, or that truth is a matter of endorsement by the powerful. This problem does not occur because mythic ideas exist, but because we don't seem able to recognize them as having their own sort of value; our errors lie in admitting mythic concepts or entities without mythic ways of thinking.

The result is a too-familiar kind of horror. A person as profoundly powerful as a United States senator can, without apparent shame, prefer Genesis to science *as a source of facts*. He can confidently put to use what he sees as factual knowledge (God said he was done trying to wipe out humanity) to further the economic profits of his supporters, even though this requires him to deny genuinely factual knowledge (that the earth's climate is rapidly and dangerously changing).[58] Myth is used as fact, and fact matters for its utility, usually of the economic sort. An even more powerful American president can decide that he prefers his own set of "facts" to those of experts, not just in the sciences, but in all things, while his staff gleefully popularizes the concept of "alternative facts."[59] The problem is certainly not facts. But it is not myths, either. It is the dual reduction of truth to fact and fact to calculable utility, set alongside the appeal of myths that no longer have any place.[60] If myths have truth, and if truth is only facts, then we needn't bother to ground facts in either reason or observation. Myth has not been overcome, but it has been misunderstood and put to misuse. Might there be happier ways of reconsidering its value?

My Stars, It's Full of Gods

> The case against Thales is that he may well have believed in an active role for a god or even many gods.
> Andrew Gregory, *The Presocratics and the Supernatural*

The story as I've told it is neither false nor (I hope) altogether uninteresting, but we can finally turn to the complicating bits, which will lead us into all the most interesting places. The Aristotelian tale of philosophical history is widely received now as *the* story of the discipline's emergence, but this opinion was less universal in antiquity. Zeno of Citium, credited with founding Stoicism in the third century BCE, associated Thales's water with Hesiod's Chaos as that which flows forth and changes form, understanding the former as the development of a line of thought rather than a serious disruption in ways of thinking. The view that Thales's concept of water as material origin was indebted to Homer's narrative of Ocean as the source of all things was fairly widespread in antiquity.[61] Both views see Thales within a tradition, rather than positioning his thought as a radical break.[62]

Thales himself gives us another kind of ground for doubting the straightforward story of philosophy's birth. While he is best known for his claim that all is water, we do know several more of his ideas. He held that the earth was not only originally made of but also still rested on water, hardly a bizarre claim for a man from the Aegean coast.[63] But he also said that magnets were ensouled. This is perhaps because they can cause iron to move—he might have assumed, like many ancient thinkers, that soul is what causes motion.[64] That explanation could help us to keep Thales safely distant from myth; perhaps he is just using the word "soul" (*psyche*) because it is available for "force that makes things move." But more damningly, both Aristotle and Diogenes tell us that Thales declared that all things are "animate and full of divinities."[65]

To save Thales's philosophical and scientific credibility, philosophers today often suppose that he was simply insisting that natural laws covered all causality; that is, that the causal principles once attributed to the gods had been brought into the world of physical things, so that all things are "full of divinities" in the sense of being themselves the source of causes. Some early commentary may support this claim by making "originary principle" or "everlasting thing" essentially synonymous with "divinity." For instance, in the fourth century CE, the heresiologist Hippolytus writes that for Thales "water is the generative principle of the universe," and the "Deity is that which has neither beginning nor end."[66] As Patricia O'Grady points out in her study of Thales, the *arche* would count as divine simply by being everlasting: in antiquity, everlastingness and divinity are quite often paired.[67]

When Aristotle comes to write his own *Physics*, he draws on the Milesian philosopher Anaximander, who is, like Thales, a material monist—though Anaximander holds that the primary substance is not water but *apeiron*, the unbounded. The *apeiron*, in Aristotle's description, "does not have a first principle, but seems to be the first principle of the rest, and to contain all things and steer all things . . . and this is divine. For it is deathless and indestructible, as Anaximander says and most of the natural philosophers."[68] Hippolytus writes that for Anaximenes, the third of the Milesian monists, "the originating principle is infinite air, out of which are generated things existing, those which have existed, and those that will be, as well as gods and divine (entities), and that the rest arise from the offspring of this," giving the material principle a divinity even prior to the gods'.[69]

Aristotle's own first principle, which sets all things into motion and is the ultimate source, is "something eternal which is both substance and actuality."[70] And then, a little surprisingly, Aristotle himself finds it entirely acceptable to consider this "something" divine; to call it, even, (a) god:

> A tradition has been handed down by the ancient thinkers of very early times, and bequeathed to posterity in the form of a myth, to the effect that these heavenly bodies are gods, and that the Divine pervades the whole of nature. The rest of their tradition has been added later in a mythological form to influence the vulgar and as a constitutional and utilitarian expedient; they say that these gods are human in shape or are like certain other animals, and make other statements consequent upon and similar to those which we have mentioned. Now if we separate these statements and accept only the first, that they supposed the primary substances to be gods, we must regard it as an inspired saying and reflect that whereas every art and philosophy has probably been repeatedly developed to the utmost and has perished again, these beliefs of theirs have been preserved as a relic of former knowledge. To this extent only, then, are the views of our forefathers and of the earliest thinkers intelligible to us.[71]

For Aristotle, then, an entirely natural but fundamental cause that lasts throughout time is what "god" has long meant.

O'Grady says firmly that whatever divinity Thales has in mind, his "hypotheses did not include the idea of gods in a religious sense. The idea of a separate divine spiritual being cannot in any clear way be associated with Thales."[72] This is a rather particular understanding of what a religious sense might be. Whitmarsh argues that the Milesian monists generally offer "a metaphorical extension of the traditional language of divinity rather than an affirmation of the gods as conventionally understood," which reminds us of "just what a vague and flexible concept a god is."[73] It is only if we take "gods" as entities and as distinct from what they create that Thales can remain a philosopher and a scientist who is untainted by poetry or myth. Distinguishing Thales and his fellow philosophers from the myth-making poets, Curd calls the world of Hesiod and Homer "god-saturated."[74] Yet a world where "all things are animate and full of divinities" certainly seems every bit as god-saturated as the cosmos that Hesiod describes.[75] Aside from efforts to recast them as scientific, Thales's claims for a divine and animate world are surprisingly often ignored. Maybe they are regarded as more common and less groundbreaking than his move toward what would become philosophy and science, but certainly they are inconvenient, and perhaps a little embarrassing. To think of gods beyond "separate divine spiritual beings" risks making Thales, for whom all things are full of divinities, theistic in some way after all. This would be uncomfortable for our origin story. It would also make that story more interesting.

There were in fact early efforts to understand explicitly the world without gods. Whitmarsh tentatively offers the honorific of first Greek atheist to Hippo of Samos, another proponent of water as first principle.[76] Aristotle is aware of Hippo, but not very impressed by him; having noted Thales's views, he adds, "I say nothing of Hippo, because no one would presume to include him in this company, in view of the paltriness of his intelligence."[77] With an equal lack of enthusiasm, Diogenes Laërtius offers us the names of "Theodorus the Atheist, who used every kind of sophistical argument,"[78] and Bion, who "would often vehemently assail belief in the gods, a taste which he had derived from Theodorus."[79] Diogenes seems to have decided that Theodorus, for all his pernicious influence, does not merit a section of his own describing his philosophy; nor does Diagoras of Melos, a poet and atheist sometimes apocryphally connected to the atomist Democritus. Whitmarsh draws upon the second-century Christian writer Athenagoras to propose, "tentatively, that Diagoras was the first person in history to self-identify in a positive way as an atheist, and . . . others like Theodorus followed him."[80]

It begins to be clear that God as a principle of nature, or divinity distributed throughout the natural world, is equally annoying to those who want their gods to be independent entities ("separate divine spiritual being[s]") and those who do not want them at all: that is, to conventional senses of theism and atheism alike.[81] It looks suspiciously like pantheism or at least panentheism. Mary-Jane Rubenstein has set forth in elegant detail the fear and outrage that such positions have historically inspired; they serve, she writes, as "limit-position[s]— marking the boundary of philosophical respectability—for thinkers of nearly every school and political persuasion."[82] Yet the divine principle of nature appears persistently in ancient, and then in medieval, philosophical texts. Aristotle, with his typical precision, may be unusually clear in his refusal of vulgar anthropomorphizing of divinity, but he is hardly alone. Thales's younger contemporary Xenophanes had declared that poets erred in giving the gods human characteristics, particularly such unsavory traits as lust and deceptiveness. He notes wryly, "Ethiopians say that their gods are snub-nosed and black; Thracians that theirs are blue-eyed and red-haired"—and further, "If horses had hands, or oxen or lions, or if they could draw with their hands and produce works as men do, the horses would draw figures of gods like horses, and oxen like oxen."[83] Though Xenophanes is sometimes called a theologian, Whitmarsh suggests that we might also consider him as lying at the origin of atheism, since he denied anthropomorphic gods.

But as theologian Denys Turner once noted, what atheism means tends to depend upon what version of theism it is rejecting.[84] If "god" means a transcendent version of an ox or a redhead, then denying that there is a transcendent version of those things is enough to declare atheism. An atheism opposing a more sophisticated theology might object less to claims about the specific forms of gods than to gods as entities at all: there is no such *thing* as a god. As Turner points out, even this attempt runs into difficulty, since this nonentity status of the divine is a starting point for many theologians as well, including such conventional Christian dogmatists as Augustine and Thomas Aquinas.[85] Likewise, Thales's divinities in all things don't exactly seem to be things themselves, nor are the senses of divinity developed in either Plato or Aristotle. The difficulty becomes still more marked with the development in late antiquity of Platonically influenced theologies that describe god(s) only by negation, much as Socrates in the *Symposium* describes the mystery of the highest beauty:

> First, it always *is* and neither comes to be nor passes away, neither waxes nor wanes. Second, it is not beautiful this way and ugly that way, nor beautiful at one time and ugly at another, nor beautiful here but ugly there. . . . Nor will the beautiful appear to him in the guise of a face or hands or anything else that belongs to the body. It will not appear to him as one idea or one kind of knowledge. It is not anywhere in another thing, . . . but itself by itself with itself, it is always one in form; and all the other beautiful things share in that, in such a way that when those others come to be or pass away, this does not become the least bit smaller or greater nor suffer any change.[86]

For many centuries, theologians have been interested in the concept of a god as other than some thing. Often contemporary atheists, perhaps unaware of this history, are inclined to view this nonentity divinity as a means of cheating, trying to have one's theism without the intellectual embarrassment that should rightly accompany it—assuming both that myths are facts and that gods are beings. If we assume neither, the world must become more interesting along with the gods.

Strange Saying

> Whether in fear or hope, this is what one expects—either a sober discourse or a drunk discourse. Sobering up or inebriation. We might even think: reason or passion, philosophy or poetry.
>
> Jean-Luc Nancy, *Intoxication*

The distinction between myth and fact has to do not only with content and justification, but with the ways in which each can be said, and the point of saying them. As I've noted, Plato is widely credited with clarifying an already extant distinction between myth and philosophy and showing the greater value of the latter.[87] But once again matters are more complex. Given the immense influence that Plato has on Wisdom-related thought in the late ancient world, it may be worthwhile to look not just at the general confusion of myth with fact, but also at the particularly Platonic ways in which that distinction fails to hold up after all—not because the two are the same, but because they work best together in presenting a complex world.

Plato is slippery. Socrates, his usual protagonist, is a famous ironist: Socrates presents himself as unknowing, asking question after question by which his interlocutors' ignorance is gradually revealed, while never admitting his own greater knowledge. Socrates, as we've noted, presents his own wisdom as no more than the knowledge of his knowledge's limits, a knowing of unknowing, of which he is more acutely aware than most. Plato's irony is less direct, but he constantly undercuts what seem to be definitive claims in his work. Not least: that reason is to be preferred to myth, prose to poetry, logic to lust, abstract universals to sensuous particulars.

As Robert Fowler points out, there are indeed several passages in Plato that suggest a distinction "between *mythos* and *logos* along the lines of falsehood vs. truth, imagination vs. reality, fictional narrative vs. logical analysis—an understanding of 'myth' which corresponds to ours in obvious ways."[88] And Plato can certainly appear hostile to myth. Rival philosophers are dismissed by the Eleatic stranger in the *Sophist* with the claim that they "appear to me to tell us a myth, as if we were children."[89] In the *Republic*, Socrates famously condemns most poetry, music, and the arts generally, urging that they be kept out of the justly run city as potentially "a corruption of the mind."[90] They are purely imitative, he declares, knowing "nothing of the reality but only the appearance," and far from improving us, they lead us to disgraceful emotional excesses. They have no place in a world, or a soul, governed by reason; "there is from of old a quarrel between philosophy and poetry," an "ancient enmity."[91] Myth fascinates, seeking engagement. Philosophy analyzes, seeking the truth. One is affectively persuasive; the other rationally demonstrative.

In this same passage, Socrates does allow an exception. "If the mimetic and dulcet poetry can show any reason for her existence in a well-governed state," he writes, "we would gladly admit her, since we ourselves are very conscious of her spell. But all the same it would be impious to betray what we believe to be the truth."[92] Judging from this passage, truth is often uncharming, but charm *may* be allowed to serve the truth, if its pleasures aren't too distracting.

As this suggests, one rational ground for the use of myth might be that it is a pleasing way to present otherwise dry points that are supported by reason.[93] That myth can be useful, though, does not mean that it might have any value that logic doesn't have on its own. Along these lines, Dexter Callender suggests that mythic truth lies "in the extent to which it *conforms* to *philosophical* discourse," where philosophy is a wholly rational endeavor.[94] False stories may contain elements of truth, like the fables we tell to children—"The fable is, taken as a whole, false, but there is truth in it also," says Socrates of a story in the *Republic*.[95] The truth is not inherent to the fable, but lies in its conformation to the truth that philosophy knows. Plato's *Protagoras*, whose titular speaker is a Sophist, suggests that the Sophistic distinction between *mythos* and *logos* is simply a matter of presentational style—mythos is an imaginative fiction that demonstrates a point, logos a rational argument to do the same.[96] One should be wary of trusting Sophists, though—Plato's Socrates is highly critical of sophistry. His chief objection is carelessness with truth; the Sophists are concerned with persuasive rhetoric, and their interest in the use of language is only about turning it to their own advantage. Language, in other words, is for them not argument but persuasion, so that it all functions as myth. They pride themselves on their ability to make any case sound like a good one; they are perfectly happy to let story or reason equally twist the truth, or rather to let the appearance of reason lead the listener away from fact—they are unconcerned with whether they reason well or badly, so long as their claims sound good.[97] They argue only to persuade; contrarily, the philosopher persuades only in order to draw the listener to reason. Spyros Orfanos marks a difference between myth and reason in which myth and sophistry come close to being the same: "Unlike myth, which requires emotional participation and ritualized recollection, logos tries to establish the truth by means of careful inquiry in a way that appeals only to the critical intelligence."[98]

But for all his love of critical rationality and all his rational criticism of persuasion, Plato's Socrates has a way of wandering into myth even as he argues against it. He employs both existing stories and novel creations, some wholly new, others combining or modifying traditional myths.[99] Even the republic from which poetry is to be ousted is actually an imperfect city—Socrates first proposes an ideal in which people live simply but without deprivation.[100] These happy people "will feast with their children, drinking of their wine thereto, garlanded and singing hymns to the gods in pleasant fellowship."[101] It is only after his conversation partner Glaucon insists on a more "luxurious city" that, with another touch of irony, Socrates constructs a society with ascetic restrictions on the arts. And the *Republic* famously ends with a long myth about the afterlife, the underworld, and lost places.[102] In fact, Catalin Partenie conservatively finds "at least fourteen" myths in Plato's dialogues.[103] Not only do these tales often follow after philosophical arguments, they may even follow a philosophical condemnation of mythology, as they do in the *Republic*. It is unlikely that Plato is inattentive to what he is doing. It does not seem too extravagant to suggest that for him and for Socrates, myth and reason are not the same, but truth may depend upon both.

This subtlety goes against what are now conventional readings of Plato. But in late antiquity, as Gregory Shaw argues, the thinkers we have come to call neo-Platonists read Plato not as an anti-materialist maker of arguments, but as someone who taught and practiced transformation. The persuasive power of myth might not be directed to making listeners believe particular propositions (as if they were facts). Instead, myths might work toward remaking the listeners themselves. For these Platonists, Plato "revealed the world as theophany," and in consequence, "they recognized divine power throughout all of nature," understanding "the supernatural [as] not elsewhere but here, in nature. Gods were everywhere."[104] The ubiquity of gods does not mean ease of identification; these neo-Platonists were fascinated by the ineffable.[105] They were less interested in making gods known than in knowing them *as* unknowable. Such a move pairs nicely both with Socrates's description of limited human wisdom and with the negative characterization of the highest beauty in the *Symposium*. It also fits nicely with the traditions of a world saturated by divine Wisdom, a world full of divinities.

In the centuries just preceding the theurgical neo-Platonists, Middle Platonism already emphasized the Pythagorean element in Platonic thought. The Pythagoreans were explicitly lovers of mystery who kept their teachings secret, but we do know that they too understood the world as god-saturated, and myth as the path to a kind of "knowing" that otherwise eludes us, a knowing of the very elusiveness of our world. Number was for them the divine and formative principle of the world, appealing to human intelligence while exceeding it. Shaw argues that for the later Platonists, too, cosmogony itself is part of the category of the unknowable.[106] Creation, that is, unfolds in a world that is itself divine and cannot be reduced to the object of human intelligence. The fact that human intelligence nonetheless finds it appealing to consider the cosmos draws knowledge into the necessarily unknown. Knowing the world as saturated with gods could hardly fail to transform the knower caught up in that world.

Plato offers us a particularly clear example of the mutual workings of reason and mythical wisdom when he sets up Socrates's speech in the *Symposium*, the speech that culminates in the elaborate negative description of Beauty and later becomes hugely influential on monotheistic thought. Before beginning his own presentation there, Socrates jestingly but sharply criticizes Agathon, the previous speaker, for praising love in flowery prose when Socrates, foolishly, had thought that they were meant to praise through facts.[107] (This is not entirely gracious of him, as Agathon is the party's host.) We might be forgiven for assuming that Socrates means to proceed by logic and reason. But having set up such an expectation, he promptly shifts into a speech he attributes to the prophet Diotima—using sly references to make sure that the reader knows that Diotima, or at least this conversation with her, is Socrates's own creation.[108] Not only does Socrates tell a story; he quite deliberately frames it as a falsehood, or more exactly as a story that is not a history.

"Her" speech begins with logical argument—in fact the same argument Socrates has just raised against Agathon—but it moves into stories about the birth of the god Eros and ends with a series of steps leading the lover through levels of love that address themselves to different kinds of beautiful objects. Diotima rather contemptuously suggests that Socrates might not be capable of climbing the whole ladder—"Even you, Socrates, could probably come to be initiated into these rites of love. But as for the purpose of these rites when they are done correctly—that is the final and highest mystery, and I don't know if you are capable of it."[109] The final point, initiation into a mystery, works through intellectual knowledge and into transformation. Despite her doubts, Diotima invites Socrates to try to keep up. Loving, which begins in the desire for a single beautiful body, culminates in the love of a mystery that she calls "this very Beauty," described, as we have seen, in a nearly incantational series of negations.[110]

Socrates has reasoned his way to a myth. Telling that myth, he has said what is beyond his own understanding; he has told us that he himself cannot rationally know what he nonetheless can tell in a story, knowing it to be true. Perhaps it is not just Socrates. In myth we tell truths we do not ourselves know, sometimes by the stories' evocative powers, sometimes by being caught up in them and thereby changed—in a way that also changes our knowing. Shaw writes of neo-Platonism that understanding it

> requires that we recover something we have lost. It requires a reevaluation of thinking itself. Jean Trouillard characterized the use of reason for Neoplatonists as profoundly different from enlightened rationality as well as from its post-modern derivatives: "The function of reason," he says, "is to reveal, in the unfolding of time, the Ineffable that inhabits it." Trouillard explains that the "reason" of Neoplatonists—rooted in unknowable darkness—reveals the world as theophany, makes audible the voices of the gods, discloses the supernatural in nature, and shares in the creation of the world.[111]

Neither reason nor myth can be discarded. Each may lead us toward the other—reason toward myth as we have seen; myth toward reason by changing us and making us aware of other ways of thinking. We might, for instance, come to see mathematics in the world. Myths lead to this peculiar kind of knowing through the workings of their non-instrumental language, their evocative storytelling, their poetic ellipticism, their repetitive liturgical rhythms. As the nineteenth-century philosopher Friedrich Nietzsche says of poetry's rhythmic force, "Rhythm is a compulsion; it engenders an unconquerable desire to yield, to join in; not only the stride of the feet but also the soul itself gives in to the beat—probably also, one inferred, the souls of the gods! . . . one cast poetry around them like a magical snare."[112] Like the late ancient Platonists, Nietzsche invokes Pythagoras, who was said to use music to heal disturbances of the soul. The Pythagorean understanding of music is complex, but it is fundamentally and sensuously numerical—deeply rational and logical. Nietzsche describes the way that music aids in the transformation that Pythagorean healing is: "When one had lost the proper tension and harmony of the soul, one had to *dance* to the beat of the singer—that was the prescription of this healing art."[113] The healing and the transformation are bodily, but in no way reducible to simple mechanical facts; they suggest a world animate and full of divinities, where reason is embodied and mobile in the manner of dance and serves both knowledge and theophany.

As the emphases on persuasion and emotional involvement, perceptual transformation, ensnaring, and even healing all indicate, the kind of knowing that one might find in myth is not separable from the kind of knower one might be. Lynne Huffer summarizes the distinction as Michel Foucault, himself much influenced by Nietzsche, makes it in the late twentieth century: "This modern relation between subjectivity and truth, where access to truth does not require self-transformation, is what Foucault calls . . . philosophy. In contrast to philosophy, Foucault names the other mode of access to truth spirituality."[114] The truth of myth, with its insistence upon cultural, emotional, or even ritual and bodily engagement, demands that we risk some transformation too. And transformation demands that we stop thinking of truth as restricted entirely to fact, as quantifiable data—especially that we do not pretend that myths are factual.[115] The role of Wisdom, after all, is to enable living well in the world that she also helps to make—to transform its seekers so that they find wisdom in creation, and their own wisdom imperfectly distinct from it. Callender finds this process at work in Torah, which "sets in motion a process that follows the mystery of linguistic communication as the means of effecting the experience of a direct confrontation," so that "ultimately the experience is mimetically reproduced *linguistically* in teaching and recitation and mimetically reproduced *ritually* in prescribed patterns of behavior and appropriate conduct."[116] Not incidentally, he develops this description with an analysis of Psalm 111, where one reads that "fear of the lord is the beginning of wisdom" ("fear" here indicates not a desire to flee but the response to awe). The Christian monk Athanasius wrote that the Psalms, uniquely among scriptures, seem to those who sing them to become their own words, and they themselves are both reflected and transformed by the singing.[117] Part of the self-transformation that wise instruction induces, then, results from language's poetic qualities. Such language does not simply inform, but instructs and changes us, making us into what might "correspond" with what nonetheless evades knowledge proper. It blurs, rather than increasing, our distinction from our worlds.

In his study of the myths in Plato's dialogues, Partenie points out that Plato sometimes "seems to interweave philosophy with myth to a degree that was not required by persuading and/or teaching a non-philosophical audience." He quotes Christopher Rowe: "On the less radical version, the idea will be that the telling of stories is a necessary adjunct to, or extension of, philosophical argument, one which recognizes our human limitations, and—perhaps—the fact that our natures combine irrational elements with the rational." But on a more radical interpretation, "the distinction between 'the philosophical' and 'the mythical' will—at one level—virtually disappear."[118] This is not because the two become either identical or muddled, but because they are working so closely together, each at the other's edge and sometimes crossing over. It is because each makes us better at the other if we can think them without reduction.

This is not only Plato's method, but that of theology and of much exegesis, which must delicately protect unknowing even as it lays out the knowable around it. In Foucault's rather wistful observation, "The modern age of the history of truth begins when knowledge itself and knowledge alone gives access to truth."[119] Our alertness to the unknowable is our access to mystery, for which logic and particularly instrumental reason have no place. In post-Enlightenment thinking, "There shall be neither mystery nor any desire to reveal mystery," write Adorno and Horkheimer sternly.[120] The very word "mystery," says Jean-Luc Nancy, "properly designates . . . that which illuminates by itself, what resorts to no reason outside itself."[121] This leaves it unprovable. What is revealed is no fact previously unknown, but (the fact) that knowledge has limits—and so, like all limited things, excesses as well.

For mythic "works and songs," writes the poetically elliptical philosopher Maurice Blanchot, "what is important is not to tell, but to tell once again and in this retelling, to tell again each time a first time." Like Wisdom, these works are always creating again. "Song is memory," Blanchot continues. "Poetry makes remembrance of what men, peoples, and gods do not yet have by way of their own memory, but in whose keeping they abide even as it is entrusted to their keeping."[122] Myths of beginnings and unendings work to reveal and remind, but not of data. The truths they tell are not facts about another place, but revelations of the deep strangeness of what is, the mystery that is not other than the world. They tell us that there are always further stories. They tell us, over and over, to begin.

2
Adam's Skin
The Strangely Bounded Primal Person

In the Beginning: Puzzles of Limitation

> Aleph is infinite; creation is a limit.
> Dan Beachy-Quick, *Of Silence and Sound*

Every beginning enters into an ongoing discussion; "every beginning is a beginning over," says Blanchot.[1] We are stubbornly drawn to beginnings, and we keep retelling the stories. Surely one of the most retold of all stories comes not just from the tales of human creation, but from the very opening lines of the book of Genesis, describing "the beginning, when God created the heavens and the earth."[2] In this beginning the earth was formless and void, and darkness was upon the face of the deep, and a divine breath moved upon the surface of the waters.[3]

Which is to say that everything is weird already. There is no form, yet there is surface; the breath, which is also a wind, ruffles the waters, complicating depth and face. Creation transforms this formless earth through a series of delimitations. First, "God separated the light from the darkness," and their alternation separates night from day, a point reiterated more specifically with the "two great lights" of the sun and moon.[4] Then "a dome in the midst of the waters" serves to "separate the waters from the waters," those of the sky above and the sea below; then the seas are gathered to separate them from earth.[5] In some translations, God "caused a division" rather than separating, marking the bounding function still more strongly.[6] The world is made by boundaries. Places are established, and things are put above and below, inside and out. Like God in the Genesis story, biblical exegetes most often see that this is good.[7] According to Philo, the strict separation of night from day, dark from light, is necessary to prevent disastrous conflict. He tells us that God "put a wall between" light and dark, "in order . . . that they might not war against one another from being continually brought in contact. . . . [The] boundaries established between them . . . separated them and prevented them from clashing together."[8] Without clearly marked boundaries, the world would risk, through ceaseless violence, a return to the chaotic formlessness that it was before the beginning. God sees that the world is good only after form has been imposed upon its chaos.

With the divisions of the world established, God sets out animal life.[9] Then comes the creation of humans, "in our image, according to our likeness," "male and female."[10] There are plants, good for the humans and the animals to eat.[11] Everything "was very good."[12] Surely all manner of things shall be well.

But beginning at the beginning never does quite work out. Just as God has rested and admired and designated the Sabbath, peopled and populated and planted the earth, Genesis begins again.[13] The second chapter presents a new account of the beginning. "This is the account of the heavens and the earth when they were created," says Genesis 2:4, and we find ourselves not hovering over deep waters, but in a desert, from which there then emerges a rising spring, a depth flowing forth onto surface.[14] Of course these are really two stories, two beginnings, because Genesis a very old and multi-sourced compilation. The elegant poetry of the first version is not first, after all; it was likely written later than the second. The second presents a rougher story whose missing details suggest its inheritance from a long oral tradition.[15] The first story gives details of the cosmic creation, but only a quick remark about humanity all at once being created in the image of the gods. The second gives us little of the detail of the first until we get to the creation of the human, where much more description about human formation is offered. In this version, Adam is created alone; he is made from earth and given God's own breath.[16] Several verses later, deciding that the man should not be lonely, God causes him to sleep deeply, takes flesh from his side, and makes a woman, who is "brought to the man."[17] In both cases, though quite differently, the human divides like the world divides, within itself or from itself.

Because the tales are fused in this long canonical text, commentators have historically attempted to consider these two stories together, to see what sense they might make in their unity. Anxious to keep the beginning a beginning, they enfold the accounts, as if to conceal the clumsy seam where the edge of one story has been tacked onto another. Usually, they argue that the second story simply adds some details to the first, particularly regarding the creation and fate of humankind.[18] But in all these readings of Genesis, however canonical or esoteric, the role of edges—of boundaries, limits, and divisions and of openings and transgressions—is crucial. The parts of creation are divided, the single lonely human is divided (or made dual at its creation), a portion of the desert is split off to make a garden.

And sometimes the divisions continue to multiply. For instance, the *Targum Pseudo-Jonathan*, a scriptural translation and exegesis from an uncertain medieval date, parallels the human to divine in a passage that divides with remarkable speed: "And the Lord created man in His Likeness: . . . In the image of the Lord He created him, with two hundred and forty and eight members, with three hundred and sixty and five nerves, and overlaid them with skin, and filled it with flesh and blood. Male and female in their bodies He created them."[19]

Here the *Targum* clearly has the first creation story in mind. Many Rabbinic texts, along with the more esoteric Gnostic and Kabbalistic commentaries, have a simple but somewhat unexpected explanation for the division of "male and female" within this first description of God making humans: they make of God, and of human perfection, something androgynous.[20] Both—perhaps, if we may be optimistic, all—genders are required in order to embody the divine image. *Leviticus Rabbah* coordinates the two stories by understanding the original androgyny as making sense of the later separation of woman from man: "When the Holy One, blessed be He, created the first man, he created him as an androgynous being. . . . When it was created, dual faces [together] were created, and it was cut, and two were made. [One] back was male, [one] back was female."[21] *Genesis Rabbah* offers a nearly identical description, and both show some intellectual indebtedness to Aristophanes's story of originary, round, two-faced humans in Plato's *Symposium*, which the rabbis might have known indirectly.[22] Aristophanes has humans split in two as a punishment for their cocky eagerness to take on the gods. The rabbinic commentaries read their God's motive more generously, as creating two humans out of one so that they can be company to each other. The second-century *Apocalypse of Adam*, from the Sethian tradition, suggests instead a punishment for early human arrogance that strongly echoes Aristophanes's version. There the first humanity is as androgynous as the first God. That first God is good, but is not identified with the creator of the world: "We resembled the great angels, for we were greater than the God who had created us and the powers that were with him, whom we did not know. Then God, the ruler of the eons and the powers, divided us in anger. Then we became two."[23]

The Valentinian tradition also puts a positive value on androgyny. The Valentinian *Gospel of Philip* even equates androgyny, as human completeness, with overcoming death: "When Eve was still with Adam, death did not exist.

When she was separated from him, death came into being. If he enters again and attains his former self, death will be no more."[24] Much later, the *Zohar*, a central text in Kabbalah, takes this tradition to a further level of abstraction. It presents the sin by which the world loses Paradise as the division of two *sefirot*, emanations of divinity that, like the gnostic Aeons, are paired male and female and are very near to an indescribable divine source.[25]

In this wide range of exegetical traditions, what is good is a conjoint multiplicity, not quite singular or plural. Divinity is not ungendered but multigendered; if not formless, nonetheless undivided and difficult to describe in any singular terms. So the role of forming, as a matter of division, is not affirmed or condemned unambiguously across traditions. Division makes the world, which in most accounts is seen to be good. Division is also a move away from being like God. Boundaries are always productive; they make and sustain the created world, whether or not that world is good. But they are not always clearly placed. The androgynous creator and the gendered humans work to make the simultaneous "male and female" in the first story and "male then female" in the second into a more nearly coherent tale.

Further problems arise even within the first version—not from "male and female," but from "let us create." What God says is not "I think I'll make some people like me," but "Let *us* make humankind in *our* image."[26] The "us" seems to suggest several gods, but as the Abrahamic faiths come to emphasize monotheism, this multiplicity becomes a problem. Christian interpreters have often been eager to read the plural as the Trinity, God speaking among its own personae. There is a wider range of readings among Jewish, especially medieval, commentators. Abraham Ibn Ezra says that the phrasing might simply indicate God speaking in the plural, as royalty does.[27] But he also considers that God may be speaking to the angels, a common reading that we find as well in the twelfth-century *Sefer Kuzari* ("God created man in the form of his angels and servants").[28] Rashi (Rabbi Schlomo Yitzchaki) suggests that this was both humble and clever of God, since the consultation kept the angels from envying the man, who was like them in form.[29] The earlier *Midrash Tanchuma* makes the divine conversation partner the Torah herself, in her Wisdom-like role as cocreator—and the Torah has some reservations about the creation of humans, reservations that God eventually overcomes.[30]

One of the most interesting responses to the plural, and to my mind one of the most productive, comes from the medieval exegete and theologian Moses ben Nachman, generally known as Nachmanides, whose work is at once traditional commentary and an early version of Kabbalah. He suggests that God collaborates with the earth. Thus the "us" of the first story is not only explained, but is made to harmonize with the creation of the human from dust in the second:

> With man He said, "let us make." That is to say, I and the earth . . . let us make man: the earth will bring forth the body from the elements as it did with beasts and the animals, as it is written (Genesis 2:7), . . . And it stated, "in our image, in our likeness," since he would be similar to both of them: in the configuration of his body, to the earth from which he was taken; and he would resemble the higher ones in his soul, which is not a body and does not die.[31]

There are several other exegeses that involve the earth in human creation. The *Gospel of Philip* declares that "Adam came from two virgins, the Spirit and the virgin earth."[32] It is not unusual for Christian exegesis to see Christ as a second Adam, but the emphasis on earth's role is intriguing and unusual. In his fourteenth-century Torah commentary, Jacob Asher offers a slightly less radical version of Nachmanides's idea: although God alone creates "out of nothing," and this is the beginning of all that exists, once other things have been created, "all the phenomena which had come into existence since that day would contribute some of their own essence to the body of man. In other words, earth itself had become a partner in G'_d's creative activities and was invited to contribute to the making of the first human being."[33] We are reminded of this collaboration with the earth in the third chapter of Genesis, where God declares the dust to be both the origin and the end of humankind.[34] And the apocryphal scripture 2 Enoch offers a complex collaboration in which God and Wisdom draw upon several different parts of the cosmos to create humans, with much of the human coming from parts of the earth. The aspects that seem to be immaterial are not strictly differentiated from those that more clearly belong to matter: "On the sixth day I commanded my wisdom to create man from seven consistencies: one, his flesh from the earth; two, his blood from the dew; three, his eyes from the sun; four, his bones from stone; five, his intelligence from the swiftness of the angels and from cloud; six, his veins and his hair from the grass of the earth; seven, his soul from my breath and from the wind."[35] The last is especially nice, as the human soul is as much of the windy world as of the godly spirit, emphasizing its nature as life and breath. The world of matter is no less important than that of the angels.

Again, the two creation stories depend upon division, though in different places—the first, as the structure of the planet and its surroundings are created; the second, as the multiplicity of humans is made. But even in that dependence, there is value given to overcoming divisions too—those between genders, between the divergent opinions of God and either the angels or the Torah, between God and Earth. Play between form and formlessness, division and chaos, separation and union all seem to be at work. Our inability to provide a stable, final determination about places and limits will be important in avoiding an anthropocentric theology. The earth's agency, especially, requires more attention.

Gathering Dust, Scattering Life

> In burial, the human body becomes a component of the earth, returned as dust to dust—inhumed, restored to humility, rendered humble.
> Robert Macfarlane, *Underland*

The boundaries of these dirt-bodies are strange indeed. Already we see that androgyny may blur gender limits and that the earth that is the material cause of the people may also have been active in their making. Earth retains its mystery, unknowable, like the chaotic waters, behind its forms. At times the edges of human skin seem fairly precise, and the human well differentiated from the rest of the cosmos. At others, however, those edges seem oddly overlaid upon and crossed with those of both the world and the divine.

Let us first return to the dust. In rabbinic exegeses, the dust for the first human is not simply gathered from one convenient spot. When the Torah objects to the creation of humans because they are sure to act badly, *Tanchuma* has God reassure her by taking the dust for humans from all over the earth. This way, even if humans do behave poorly, no part of the world can refuse to take them in, because some part of them will belong there. *Genesis Rabbah*'s explanation for the dust-gathering is offered in answer to a question from the Torah as well. The Torah asks, "But why did God create man from the four corners of the earth, and not from the dust of one single spot?" The answer here too is that a human body must be received by the earth when it dies; if it is drawn from the dust of all the earth, no place will reject it.[36] In a particularly nice addition, this text specifically includes in the human mix the dust from the site of the sanctuary of atonement. Because humans can always atone, mercy is always possible, indeed is built in as a possibility. Because of this, humans can endure.[37] The commentary by Rashi approvingly cites and agrees with both texts. From the beginning, the human is made to have a place at the end.[38] What's more, both the dust-made human and the dusty world need the possibility of mercy granted by the sanctuary. Without compassion, as the *Zohar* points out, "the world would not exist for a moment in the face of harsh judgments aroused every day."[39]

The *Chronicles of Jerahmeel* suggest that the earth has grounds for wanting to refuse to make humans. Wishing to make a human, God asks the angel Gabriel to gather "dust from the four corners of the earth."[40] But the earth refuses to let Gabriel take its dust, because "I am destined to become a curse, and to be cursed through man, and if God Himself does not take the dust from me, no one else shall ever do so."[41] (The earth has in mind the punishment that God imposes on the humans before expelling them from Paradise for their disobedience, in which he declares that they will be returned to the dust from which they came.) God therefore gathers the dust himself: "He stretched forth His hand, took of the dust, and created therewith the first man on the sixth day."[42]

In all these accounts of gathered dust, the human extends, in its curiously compact way, over the entire surface of the world; or the world gathers, from all its edges, into the human. We can see one of the ways in which boundaries become unstable and strange. The human body is both of and distinct from God, whose breath it breathes, and is both of and distinguished from earth, whose substance has formed it. It is hard to think of better figures of mingling than dust and breath: clouds of dust or gusts of breath cannot come into contact and remain distinct. Already the human body, given such pride of place, is entangled with the divine likeness and image—which would certainly let it hold on to that pride—but also with all the world's dirt, which disperses the dust throughout that divine image as well. The earth is not passive; in many of these accounts, it plays actively into creating the human body, or has a say in whether the dust can be gathered in the first place, and whether it will accept the dust upon the person's death. The dusty earth, co-making and accepting, making space for, appears not passive but intriguingly wise—it is even essential to the possibility of mercy. Without appreciating its mysteries, the human fails to know itself.

Though they place less emphasis on dustiness, both Valentinian and proto-orthodox Christianities may insist on a thoroughgoing mutuality of spirit and flesh. According to the Valentinian teacher Theodotus, matter taken from the earth makes a material, irrational soul "consubstantial with that of the beasts." This does not sound so good for materiality, especially when the Creator then breathes into the human "something consubstantial with himself."[43] But "man is in man, 'psychic' in 'earthly,' not consisting as part to part but united as whole to whole by God's unspeakable power. . . . The divine soul . . . is hidden in the flesh . . . —the material soul which is the body of the divine soul."[44] The "material soul" is the body of the "divine soul," and the "divine soul" is "hidden in the flesh," joined whole to whole. The bones themselves give us the poetic image of spirit: "So Wisdom first put forth a spiritual seed which was in Adam that it might be 'the bone,' the reasonable and heavenly soul which is not empty but full of spiritual marrow."[45] In this, the Valentinian fragments draw near to the rabbinic exegeses of cooperation between earth and God. Both the precise rabbinic readings and these mythopoetic texts emphasize the inseparability of spirit and flesh, each lending itself to the other. They warn us against the misreading of mystery as simple interiority.

Tertullian's emphatic if rather surly defense of flesh is especially hostile to the Valentinians, with criticisms he may have borrowed from Irenaeus. He joins them, though, in considering flesh as ensouled throughout. He begins by taking literally the idea of a body molded from dust. God's touch, he says, was already enough to honor and transform clay (which is dust that can be formed). The honor goes further, since Adam is made as a model for what Christ will be, when humankind needs a new beginning. "For, whatever was the form and expression which was then given to the clay (by the Creator) Christ was in His thoughts as one day to become man, because the Word, too, was to be both clay and flesh, even as the earth was then. . . . Thus, that clay which was even then putting on the image of Christ, who was to come in the flesh, was not only the work, but also the pledge and surety, of God."[46] By receiving form, the clay becomes something more than clay. But it might already have been something more—the Word will become "flesh, even as the earth was then." The breath of God transforms the gathered earth according to earth's own potential, as clay would be transformed into earthenware by heat.[47] "You have both clay glorious from God's hand, and flesh more glorious from God's breathing: and by this breathing the flesh at the same time laid aside the rudiments of clay and took upon it the adornments of soul."[48]

Sources across quite a range of traditions, then, trouble the boundedness by which creation is made by intermixing the human with the rest of the material world, or even giving the dust of that world priority or agency in human creation. Participation in divinity does not sort the human-dust out from the rest of the world. Forms and divisions, redistributions and unities, divinity and ensoulment all become hard to place and restrict in this earth of intermixed bodies. The impossible body in the image of god-and-earth, god-and-wisdom, god-and-text shimmers at its edges between distinct and undivided. Grappling with this paradoxical placement, other exegetes suggest a doubling of human and world, a mapping that both succeeds and fails. Its failure may tell us more than its success.

Microcosmic, Macrohuman: The Size of the Edges

> The dance along the artery
> The circulation of the lymph
> Are figured in the drift of stars
>
> T. S. Eliot, "Burnt Norton"

The rabbinic commentaries often make much of the sheer size of the human who is gathered from the dust of the world: "He stood upon his feet, and was in the likeness of God; his height extended from the east to the west"; he is made "extending over the whole world . . . from the one end of heaven unto the other."[49] Here the Adam of Genesis is coextensive with creation, less a microcosm than another reading of the cosmos. The idea of human as microcosm is ancient and widespread—the secretive Pythagoreans seem to have been among its proponents.[50] Its antiquity does nothing to simplify it, as Ibn Ezra succinctly points out: "Now the human body is like a microcosm. May G-d be blessed, Who began with the great and concluded with the small. . . . Now Hashem is one, and He creates all, and He is all, and I cannot expound this further."[51] The latter part of this statement complicates the first; the human is not merely a small copy of the large cosmos—which would already be fascinating—but both are also the one who creates them, and that one creates itself into all without ceasing to be one. There is a limit to exposition of this matter; it cannot be gathered into a comprehensive or comprehensible whole.

Though Adam is strangely distributed in rabbinic commentary, Adam as microcosm appears especially in Kabbalah and in gnostic texts. The microcosmic human is not always identical with the human who will later come to trouble in the garden of Paradise. The Aramaic term *Adam Qadma'ah* (generally rendered in English "Adam Kadmon") appears in the *Zohar* to name the primal human, and a similar concept emerges in Sethian Gnosticism. This primal human is perfect, and enormous—sometimes itself cosmic in scale, so that it is hard to say which is the image, human or world; or if the cosmos might be human-shaped, or the human cosmic-shaped, both in the image of the formless One. Often Adam Kadmon is represented as a mediating reality between the first source and our material realm. The first ordinary human may be a secondary image of this perfect version or a diminution of it by sin. That it shares in the primal form, however, makes it microcosmic as well.

The parallel between human and cosmos is sometimes laid out as a series of similes. In one midrash, the development of the world is compared to that of an embryo: "A child starts to grow at the navel and then develops in all directions, and so the Holy One, blessed by He, began the creation of His world at the foundation stone, and built the world upon it. . . . The formation of the embryo is like the formation of the world, for just as the embryo is formed in the womb of its mother, so too the world was formed at the foundation stone."[52] Cosmogony parallels fetal development, and both set limits to make form out of formlessness.

More often, though, the simile is reversed: we read not that the world is like a human body, but that a human(like) body is cosmic. These descriptions can be quite elaborate. In *The Fathers according to Rabbi Nathan*—a rabbinic text that is often considered Wisdom literature—we read, "Man is equivalent to created world."[53] Discussion and citation of other rabbinic sources follow,[54] after which "Rabbi Yose the Galilean" lists thirty distinct parallels between the primal man and the rest of creation. Some are fairly self-evident, such as the forests of hairs on the head; some are rather more obscure, such as the walls of the lips and stars of the cheeks; still others are downright odd, such as the counselors of the kidneys, or the angel of death of the heels.[55] In 2 Enoch Adam's very name is "appointed . . . from the four component parts, from east, from west, from south, from north, and I appointed for him four special stars, and I called his name Adam."[56] Then the earth, dew, sun, stone, cloud, grass, and wind are joined by the stars to map Adam onto the cosmos. This tale risks recentering the world on humankind, but once more the placement of both is too strange to promote such an outcome. As we saw in the dust, both gathering and distribution decenter the same human that they make.

The Kabbalist *Sefer Temunah* emphasizes the corporeality of the divine image: "The *Sefirot* [divine emanations] which are the image of man—for man is a microcosm according to 'Let us make man in our image, in our likeness'—are seven Forms, and the soul is in the body and is the hidden light which is in his head. For in [the body] is the mystery of the 'small image.'"[57] The human is microcosmic by virtue of being in the divine image, so the cosmos of which it is a micro-version must itself be a divine image. "The human being is in the image of a small world," Ibn Ezra writes, so that understanding the human and soul grants some understanding of the upper worlds as well.[58]

A small world, says Ibn Ezra, as if there might be more—and Kabbalah reads the body as the image of a whole multitude of cosmoses.[59] In Isaac Luria's early modern cosmology, as his student Chaim Vital records it, the perfect first-created human and the cosmos both map in complicated detail onto the emanations of the divine, where each emanation, each of the *sefirot* paired with another, is an aspect of the singular source. (This much is common to Kabbalah broadly.) Vital describes this mapping recursively:

> It is clear and obvious that many kinds of worlds were emanated, created, formed, and made—thousands and millions. . . . Now, each and every world has its own individual ten *sefirot*. Each and every individual *sefirah* that is within each and every world includes ten individual *sefirot*. All of them are in the form of concentric circles without end or number. All of them are like the layers [or "skins"] of onions, one within the other.[60]

Though we might debate its clarity and obviousness, this passage gives us a sense of recursive nestling, worlds within worlds—"without end or number," skin within skin. If each *sefirah* includes within itself the ten *sefirot*, then so too must each that is included within in turn include its own set of ten. This nesting of spheres is the circular aspect of emanation from the divine. If the macrocosm has a mirroring microcosm, then to mirror properly, each microcosm must have a still more micro-version. But, says Vital, there is a linear aspect to emanation as well, "via three lines like a man (*adam*) having a head, arms, legs, body, and feet."[61] And now things get properly complicated:

> The aforementioned circles spread via this aforementioned line that had spread downwards. This line did also spread in a straight line downwards from the top of the highest roof of the highest circle to the lowest end of all the circles. Its length is composed of ten *sefirot* in the image of a straight tall upright man composed of 248 members drawn in the figure of three lines: right, left, and middle, and composed of the ten *sefirot* in general. Each and every *sefirah* is infinitely split into ten sefirot. . . . Now, this second aspect is allied with the image of Elohim.[62]

All this, Vital tells us, is what is alluded to when God (*Elohim*) declares that they shall "make man in our image." The recursion of nested circles meets the fractal division of lines, mapping the infinite in the finite images of the human body and the world body. The human in the image of the divine is the image along which the divine emanates, and the worlds are shaped like the human who is shaped like them. This way of thinking the microcosm is not widespread. It does, however, take seriously a point that any microcosmos implies: if the macrocosm is mirrored in the microcosmic image, accurate imagery requires that the small must be mirrored again in the smaller.

In its microcosmic character, the human must also mirror mystery, and this bafflement may be one way to do so. The *Zohar* emphasizes the revelation, in the human, of the hidden source that is the original of both the human and the cosmos as images. That source is so necessarily hidden that even discussion of it among the rabbis must be approached with great caution. "Now, if in mundane matters secrecy is necessary," the text cautions, "how much more so in the most mysterious matters of the Ancient of Ancients, which are not transmitted even to supernal angels!"[63] When mystery is revealed, it is revealed as mysterious.

The story of the concealed source is told in a long series of complicated images that give more detail to the cosmic body. At the beginning, "Before the Ancient of Ancients, Concealed of the Concealed, had prepared adornments of the King and crowns of crowns, there was neither beginning nor end. He engraved and gauged within himself, and spread before Himself one curtain, in which he engraved and gauged kings, but his adornments did not endure."[64] The concealed one continues to create behind the curtain. The way that he comes to be known is indirect. "The Ancient of Ancients, Hidden of the Hidden, was arrayed and prepared—that is, He existed and did not exist, did not actually exist yet was arrayed. No one knows Him, for He is more ancient than the ancients, but in His arrayal He is known . . . —known by His signs yet unknown."[65] In this arrayal or setting forth, in the only ways that he can be made known, the Concealed or Hidden One generates light and worlds. The description of this generation is in bodily form and in astonishing detail. It begins with the skull. "In the skull dwell 120 million worlds, moving with it, supported by it." The luminous whiteness of the skull, harboring Wisdom safely within it, radiates "in thirteen directions."[66] The description continues, taking in the hair, forehead, eyes, nose, and beard, to create an image of the unimaginable.[67] As an image of the immeasurable and uncontainable, number multiplies. We read, for example, of the hair: "From the skull of the head hang ten billion, seven thousand five hundred locks of hair. . . . Every single lock has 410 strands of hair, . . . and every single strand glistens in 410 worlds, and every single world is concealed and hidden away, unknown to anyone but Him, glowing in 720 directions."[68]

The source itself is beyond form and beyond knowing. Once arrayed, subject at least to the measure of number, it can be somewhat and indirectly known, and it is known as body-and-world. In knowing one layer—the human, the world, the creator—one perceives the others as well, however dimly. Revelation is not of facts, but of the transformative knowing of imagery. In each of these perceptions there is also a perception of the limits of perception before the illimitable, whether we imagine the infinite as vast beyond knowing or endlessly smaller. Knowledge is made to acknowledge the unknowable. Everything seen—or heard, or known by any sense—hints at the unknown "one."

Another strangely multiple relation of microcosm to human-like form appears in the *Secret Book of John*, one of the best-known Sethian texts. Though we do not find recursive worlds, we do find levels and layers. Here the primal human is perfect, explains Stevan Davies: "The whole of the realm of Providence, which is the pleroma, is defined as the First Man. . . . The whole pleroma will be displayed later. . . . The First Man [is] the primordial perfection of which humanity is a reflection."[69] Human perfection is at once multiple—the First Man is also the many pairs within the Pleroma, who in their pairing mirror the androgynous divine source—and singularly embodied; it is known only as it is imperfectly reembodied and reflected in the first human-as-we-know-it.

The ordinary human copies of this prototype, beginning with Adam, are created by lower powers trying to be gods themselves. These lesser creators make an imperfect human. To embody the First Man, the "seven Powers" created psyches of various aspects of the body: bone, sinew, flesh, marrow, blood, skin, and hair. A "host of demons" (or, in other translations, angels) then "took these substances from the powers to create the limbs and the body itself. They put the parts together and coordinated them."[70] The parallel to the more positive idea of human as divine image has to be taken up with caution here, since these demons are not lovingly portrayed. But even here the body-soul difference is not as distinct as more modern thought would have it: the body is gathered and coordinated from a series of *psyches*.

Coherent though the First Man might be, an unnerving formlessness flows behind the copy.[71] As the demons work to form the body, "Their mother stands among them . . . /She is unlimited./She mixes with all of them./She is matter/and they are nourished by her."[72] By this indiscriminate mixing, unlimited Matter also unifies and coordinates the scattered parts of the psychic human; as mother, she makes the work of her forming, gathering, coordinating offspring possible. Though it will be up to higher forces to give the human a spirit (*pneuma*), in total, 365 demons "work together to complete, part by part, the psychic and the material body."[73] Matter mixes with all of them, unlimited by form—but necessary to form's instantiation. What seems at first an affirmation of the superiority of pure immaterial form also affirms the dependence of that form upon equally unknowable matter. Indeed, it may be matter that proves the more epistemologically elusive. Matter is unnervingly shapeless, alarmingly indiscriminate, and powerful: every aspect of the human, whether we might consider it body or soul, is mixed with her. As a potentially revelatory image of the greater cosmos, the human is also the presentation of the unknowable formlessness, matter made into forms as the only way that it can be perceived. Matter may be lower, indiscriminate, and imitative, but without it there is no revelation.

The human as divine image and microcosm appears with a new complication in an untitled fourth-century text in which, as David Brakke writes, "Bodies and boundaries multiply at a dizzying pace."[74] Brakke places this work "outside the boundary of Gnostic Christianity," but notes that its author "has drawn on several currents, especially in Valentinian thought, to use the human body as the guiding metaphor for what human beings can and cannot know of the divine."[75] Here too the work of shaping is part of the work of mirroring. The text describes "numerous places within the divine realm—for example, the height, the holy fullness, the immeasurable deep, the god-bearing or -producing land," which are also bounded spaces both with or without openings; and it describes a strange composite perfect human, who is also a city, which is the image of the divine realm of the Pleroma.[76] The treatise declares, "The Father took their entire image and made it a city or a human being, and he depicted those entireties on him, that is, all these powers.... And the Father took the glory and made it a garment external to the human being (266.20–267.2)."[77] As Brakke clarifies, the characteristics of the city tell us of the heavenly realm; the characteristics of the body that the city also is tell us of divine bodies forming a divine realm too; the glory given to the Father is also given to the human who is the city. The glory is wrapped around the city-human, in the manner of both a skin and a city wall.

Everything here is in some way incomprehensible, not least in relations of containment and dependence. The very limitation of the city-body functions to sustain both existence and creation. To limit—to make boundaries—is to participate in divine creativity. But those boundaries are established only in such strange ways and places that the usual function of marking inside and out, greater and lesser, image and copy becomes its own enigma.

Finally, the unmappable extension can be temporal as well as spatial. Alexander Altmann notes that in Kabbalah, the image of the microcosm is unfolded "on the threefold level of *mundus-annus-homo.*" He cites the early Kabbalist Isaac the Blind, for whom the human is "composed of the supernal and the lower [forces], and he belongs to the world, the year, and the soul. For all that is in the world is in the year, and all that is in the world and in the year is in the soul."[78] In Kabbalistic commentaries such as the *Sefer Ha-Bahir* and the *Tiqqune Ha-Zohar*, the limbs of the human body are an image not only of the lower *sefirot*, but of the days of creation as well, each numbering six.[79] Time too appears both to nest and to divide.

The macro- and the microcosmic, the spatial and the temporal, the here and now and the eternal, are not so much contained within the human as they and the human are contained within one another, emphasizing the impossibility of a still and orderly conception that could hold them distinct. The strange play of inside and out reminds us that we cannot neatly differentiate them, not even metaphorically. Each set of boundaries is the image of another, at times so exactly overlaid as to be beyond distinguishing, but at others—or even at the same time(s)—beyond locating. Limit conceals unlimitedness, forming matter into a revelation of the formless unknown. The matter that presents form shocks us with its own incomprehensibility. The form that it presents is that of the unformed origin, but also a hint of matter's own unknowable malleability. Just as the persistently unplaceable dust kept us unable neatly to distinguish human from earth or to know earth as an object, so too the persistently unplaceable boundaries of the microcosm keep us from setting any altogether apart from the others. So far, this unplaceableness has appeared in space, in time, and in affective response as a striving at the edges. It appears in epistemology, too, where light and vision seem to curl back on themselves. Humans, who are not the central stuff of the cosmos, are decentered as knowers, too.

The Light That Sees the Eye That Shines

We are in many ways the universe looking back on itself.
David Baker, "Big History, Critical Thinking, and Transdisciplinarity"

The *Derech Eretz Zutta* says, "This world resembles the eyeball of a man. The white is the ocean that surrounds the whole land; the black is the world; the circle in the black is Jerusalem, and the image (the pupil) in the circle is the Temple."[80] Upon what could the whole of the world set its gaze, and how could it do so?

Perhaps, as several esoteric traditions suggest, it looks at itself, and by its own light. In the *Secret Book of John*, the intersections of light and what light shows, and of knower and known, develop through the story of human creation. Here light hides as it shines in order that a human image of divinity may enter the world. The process begins with a light that somehow appears in a world without matter: the Father "illuminated the waters above the world of matter,/His image shown in those waters./All the demons and the first ruler together gazed up/toward the underside of the newly shining waters./Through that light they saw the Image in the waters."[81] The image is not quite made of light, but of light in water, which is itself between materiality and the immaterial and, as flowing, between formed and formless, as well. The Father illuminates the waters so that a true image shows in them, an image visible from the material realm. The light is both the illuminated image and the glow that illuminates.

The rulers of the material world attempt to make an image that will also create its own illumination—not a light that takes form, but a form that makes light. They base their effort on their perception of the divine image shining in the water above them. "Let's create a man according to the image of God," they tell one another, "And our own likeness,/so that his image will illuminate us!"[82] They realize, a little too late, that the body-image of the Father, the image that they are using as their model, shines with light that is not theirs to take. Discouraged, they leave their dim creation to the depths of materiality—to the unlimited mother who takes all forms.[83]

But contrary to their intent, the created body, which will be Adam, is not abandoned there. Instead, "light-filled Epinoia," the wise thought that emanates from the first source, enters into the body as the body's helper.[84] She personifies the capacity for reflection—light returned to itself—that will enable Adam to recall his roots in divinity.[85] That is, her light will enable him to see his own. Epinoia will illuminate the "shadow of light" dwelling in Adam—a strange term that makes light and shadow inseparable.[86] The material man is now inwardly luminous, and because of this he can perceive light. His perception, which is granted by Epinoia, is light gazing at light like an eyeball that stares at itself. Her role as Wisdom serves as a necessary and redemptive reminder.

As they attempted to replicate the perfect First Man, so too the rulers of the world try to create a woman in the image of the perfect Epinoia. Once more they succeed only in making an inanimate material vessel. As she had previously done with Adam, the real Epinoia now enters this second vessel, newly concealing herself from the rulers. Worse still, from the rulers' perspective, she is thereby revealed to the man, who because of Epinoia's previous presence in him can recognize her luminosity. "Adam saw the woman standing next to him. The light-filled Epinoia immediately appeared to him.... And he recognized his own counterpart."[87]

In this sequence, human creation echoes cosmogony, as we earlier saw cosmogony echoing human form. Stevan Davies explains the cosmogony of the *Secret Book of John*. In the beginning, "The One gazed into its own light and saw itself reflected there. His self-aware thought came into being, appearing to him in the effulgence of his light. She [the thought] stood before him . . . prior to everything, . . . her light reflects his light. She is from his image in his light."[88] The world begins when the divine light-eye gazes into and out onto itself, itself the light that illuminates what it sees. Even self-awareness requires this movement out of itself, the thought that stands before the thinker. From the light of the created wisdom "within," Adam can recognize the first light, as the light recognized itself, and formed the first thought that created wisdom is. Adam can only recognize the light because it is also his own—as the neo-Platonists say, the eye can see because it resembles the sun.[89] But the sun by itself is not enough for seeing. The light must go outside itself to see itself. It is only when it is formed by matter that it is visible.

Beginnings multiply. "Let there be light," the first imperative from Genesis, becomes the beginning in which the One gazes into its own light and sees its reflection—though there can be no thing to see yet, just light upon light.[90] In the beginning-again, Epinoia shows her luminous self to Adam, thus showing Adam to himself as well. But she shows him herself from inside him, and then shows him the light of himself (which she is) from another body. Light and thought must double or exteriorize themselves to be visible to or know themselves. Epinoia takes material form, and the origin can know its first, bright thought.

Altmann argues that midrashic sources will also read Adam by this gnostic light. In the rabbinic midrashim, Adam both shines and sees by the light that is alternately his possession or himself. *Leviticus Rabbah* tells us that even "the apple of Adam's heel outshone the globe of the sun; how much more so the brightness of his face."[91] (There is a pleasing resonance here with a fragment from Heraclitus, which says, "The sun by its nature is the width of a human foot, not exceeding in size the limits of its width.")[92] The Talmudic *Bava Batra* tells us Adam is extremely beautiful and his heels as bright as not one but two suns.[93] Altmann notes that the rabbinic concept of the primordial light is intimately connected to the idea of the primordial man.[94] In fact, "it appears that the primordial light which was created on the first day is identical with the lustre of Adam."[95] The perfect Light-Human sees its own extension, from one end of the world to the other, by the light that it ambiguously has and is. And after all, if that human coincides with the world, from one end to another, there is nothing to see but itself. The world itself must self-know.

The concept of the luminous perfect sort-of-human is developed in Kabbalah, where Lurianic versions are once more especially intricate. Chaim Vital writes of Adam Kadmon,

> Now, many lights emerged from him and shine outside of him. Some of them developed from his brain. Some of them are from the skull. Some of them are from his eyes. Some of them are from his ears. Some of them are from his nose. Some of them are from his mouth. Some of them are from his forehead. Some of them surround his body, which is the aspect of his seven lower [*sefirot*]. Many shining lights surround them and they depend on them.[96]

Both emerging and surrounding, coming from and depending on, this illumination is hard to place in relation to what it makes visible. The lights surround; the lights are surrounded. Serving as a boundary to display the image of divinity, the skull can only show the unshowable and can only conceal by illuminating. The brightness conceals the mystery, hides it within the light—but it is light that allows us to see. Light, which radiates out, becomes a boundary, too. The boundary shows us concealment.

On some rabbinic readings, Genesis tells us that the first humans, created to live in Paradise, were purely innocent and divinely radiant, with nothing to hide (or to show), no need for darkness and shadow. They were safe in the light.

The image of light as safety for the pure appears again in the *Secret Book of John*, though not in regard to Adam: Noah and his family are spared from the worldwide flood by going "into hiding within a cloud of light!/Noah knew his own authority/and that of the Light Being who illuminated them."[97] In fact, the image of a protective, enwrapping, or concealing light is remarkably widespread; in the *Gospel of Philip*, light is donned sacramentally and becomes protection from lower powers.[98]

Light, which we customarily connect to revelation, is valued as concealment.[99] In several readings of Genesis, this concealment is what humans lose when they disobey their God. The prelapsarian humans are pictured as being wrapped or clothed in light—a description that is also sometimes applied to the creator.[100] The light may even be of their own bodies. In the *Genesis Rabbah*, "The garments made by God were not of skin, but of light,"[101] suggesting a contrast with the leather garments in which the humans are clothed when they are sent out of Paradise.[102] Here the "garment" is not distinct from the body but is the outermost layer of it. And as a consequence of sin, Adam loses "his brilliance"—not his intelligence, but his light.[103] Isaac Luria plays on a small diacritical difference to also suggest that the human garment of light becomes one of skin: "After Adam sinned, his clothes changed from light [aleph-vav-resh] to leather [ayin-vav-resh]."[104]

Even before they are sent out into their leather-bound lives, the humans lose the protection of the light. When they eat from the Tree of Knowledge, they come to certain realizations, notably that they are naked. But this knowledge is not illumination; instead, it leads the humans to try to hide in the shadow.[105] Unsurprisingly, God finds them out. In the version from the *Targum Pseudo-Jonathan*, "the Lord God called to Adam, and said to him, Is not all the world which I have made manifest before Me; the darkness as the light?"[106] Having found the humans, manifest in the dark, their creator exiles them from Paradise.[107]

When Eve and Adam acquire knowledge, they simultaneously acquire shame. Eyes opened, they see that they are naked and diminished. They are no longer the light that illuminates itself, and that is what it sees; they are no longer wrapped in light and safe there. This time, division is not a good. They have become bounded and separate beings who cannot wholly see themselves, whose world is small and constrained, and whose attempts to hide in the dark, because they are no longer luminous enough to hide in light, fail when divine light finds them anyway. By their sinfulness, humans are rendered more clearly distinct from the world. To be set apart from the world is a fallen, diminished form of human life.[108] To be only human is to be less, and to know less expansively and less clearly.

These cosmic, creative bodies seem to be both visible and not, both light and unseeable within the light itself. The visible is light's enwrapping, its skin; equally, it is the illuminated "interior" that light shapes and enfolds. It would be easier to assume that the light within visible things is inside as if in a container, or on the other hand that there is a clearly defined shape that glows a bit at its edges against a darker background. But there is no proper, consistent containment at work here. It is only in the play between same and other that light can illuminate itself, that the eye can see itself and can shine its own light, that the light can look at its own self. The eye that sees itself must place itself outside for seeing, and so it can no longer be itself alone. In this illumination mystery glows as brightly as knowledge.

Or, as the *Zohar* tells us, "This is the light that the Blessed Holy One created at first./It is the light of the eye./It is the light that the Blessed Holy One showed the first Adam;/with it he saw from one end of the world to the other."[109] By the light of the eye, we cannot tell seer from seen, nor light from what light illuminates. World-eye-microcosm-God-person gazes impossibly upon itself, itself that is nonetheless not self-contained. Whatever knowledge this is, it is not comprehension. Whatever world this light creates, the edges of its form will not stay still against the dark.

It is hard to tell where or to what the light belongs—to God-cosmos or to human-world; to creativity, ontology, epistemology, or soteriology. Illuminated from within, the human gaze becomes reflection, setting it perilously proximate to the gaze of God: they gaze with the same eye.[110] This is a double problem—a problem for the desire to keep the creator strictly transcendent, and a contrary-seeming problem for our hope to avoid an anthropocentric world.[111] We find a similar issue, in more narrowly epistemological terms, in the widespread gnostic idea that self-knowing is essential not because the individuated self is important, but because, as Marvin Meyer writes, "To attain this knowledge—to become a gnostic—is to know oneself, god, and everything. . . . According to many of these sacred texts, to know oneself truly is to attain this mystical knowledge, and to attain this mystical knowledge is to know oneself truly."[112]

To say that knowing oneself is knowing everything could, of course, glorify the self. But it could, as well, lose the security of the self along with that of transcendent divinity. To know oneself could mean something other than arrogating the cosmos to that self. It could mean knowing the human as particles of the same sacred dust that drifts everywhere—or as the not quite placeable light that illuminates the dust motes. The deep interiority that human self-knowledge seems to be, a complement to its subjective ability to perceive a world of objects, becomes instead a knowing and a knownness so omnidirectional that they are lit from everywhere and nowhere by a light that shines upon itself—and thus reflects upon its own limit, the unknowable blind spot at its source. Knowledge illuminates its own unknowing. Or, to cite the *Gospel of Philip*, "Some things are hidden by the visible."[113]

In twentieth-century philosophy, phenomenologist Maurice Merleau-Ponty, considering what he is among the first to call the flesh of the world, develops the point that we cannot see without being visible. This necessary reversibility will be taken up by new materialists to undermine the strict binaries of subject and object, action and passion.[114] As seers, we can be seen in ways that we cannot see. To know ourselves must be to know how shifting the world's boundaries are, our own included. This is not a simple contradiction to the idea of an eye that sees itself. When the eye sees and the light illuminates itself, they illuminate both knowing and unknowing, both seeing and unseeing. The origin can see only by thinking, desiring, moving outside itself: even there, vision cannot turn perfectly upon itself. There is always a blind spot in knowledge. The myths and exegeses of bodies and light gathered here turn their gaze upon a divinity that is as visible and invisible as flesh—visible and invisible because it cannot be taken apart from flesh. The very condition of the visible tells us that not all can be shown—even in all that we see. Concealment and revelation, the light dimmed and sparking, hidden in bodies and recognized through them, cannot be told without one another. "This is the way it is," the *Gospel of Philip* concludes of the revelation, the true knowing, given by the sacraments. "It is . . . hidden not in darkness and night but hidden in perfect day and holy light."[115]

One way to say that mystery, the mystery of matter that gives itself for thinking of and for being, of knowing that curls back in reflection, is that the dust is light itself. In *Black Fire on White Fire*, discussing cosmogony in Kabbalah, Betty Rojtman describes the changes in the light, the light that flows out of namelessness to become creation. "These rays of light as a remainder, as memory of an absence [of the contracted, withdrawn first source], as puddles of light, gradually grow thicker, from world to world, in order to make themselves into dust, a material substance."[116] Having become dust, the light can be gathered into the primal human-cosmos-time, the image of the Concealed—the light that shines on its own incomprehensible shape. Matter itself, the dust that the world is, remembers making—and as the memory of an absence, recalls the source as immemorial, inconceivable. The dust is the light, gathered and scattered, known as unknowable, distinct in its unity, knowing and illuminating itself.

Somehow, through the concept of the material body as an image (however poorly copied), through the breath that moves in and out of bodies (however unplaceable in its motion), divinity is revealed. It is elusive and slippery and dusty and drifting. The revelation offers not a subject's knowledge of an object, but a recursive and imperfect self-knowing that must include knowing knowledge's limitation. Light radiates, coalescing into matter in the image of the body at the origin of its own radiance. Revelation is transformative, but the new forms continue to transit; the epiphany is never static. What seems at first to be a simple containment, the divine light or spirit concealed or at least held inside the flesh, the flesh compactly gathered inside the skin, is complicated repeatedly both by the language of cosmos and image and by the unlocatability of that light or spirit or God. The anthropocentric mapping of human onto cosmos disintegrates through the very act of gathering. The edges made by separation and division only let us see that the illimitable is. That these tales and exegeses are complicated, and not entirely consistent even within themselves, is evident.[117] They are also deeply paradoxical, and as so often happens in theological discussions, the paradoxes are revelatory. They highlight the strangeness already implicit in Genesis: that the lowness of dust is the breadth of the earth, is the image of the divinity as the cosmos; that the knowledge of self and of earth and of the divine cannot be just knowledge at all, but must, in the midst of the perceiving world, be wise and desirous enough to see the mystery there, to know oneself in the drift of the world, lit by the dust.

3
Limitless Bounding
The Valentinian Body of Christ

A Saving Body

> For to the people the Messiah was an expected good, which the prophets had foretold.
>
> Origen, *Commentary on the Gospel of John*

The first set of divisions in Genesis, of things from other things, is the necessary condition of things' existence, and is canonically regarded as very good. The division of humans from divinity, and arguably from the rest of the world, is not so good, and these negative divisions generate unhappiness in a damaged world. Damage demands repair. To mend such a disruptive break, there must be a new beginning. The work of salvation offers this second chance, and often it begins with the arrival of a messiah.

In the twelfth century, Moses Maimonides offers a nice summation of what becomes the consensus after rabbinic Judaism: "The King Messiah will arise and re-establish the monarchy of David as it was in former times. He will build the Sanctuary and gather in the dispersed of Israel. . . . The first Messiah was David who saved Israel from her adversities. The final Messiah will be from his son and will deliver Israel from the hands of the descendants of Esau."[1] In fact, Maimonides affirms messianism as the twelfth of his thirteen principles of Jewish faith, reminding his readers not to be impatient as they wait.[2] The messiah is a figure who might have particular gifts, profound knowledge, or a royal heritage. In all these ways, he (the Jewish and Christian messiahs seem always to be male) is remarkable.[3] He certainly has a special role in history, and often a special relation or access to divinity and wisdom. His arrival may portend not just a political restoration, but a transformation in the very nature of the world and what it means to live.

There is more variation, though, in arguments about the messiah's body, and whether it too is something exceptional. Maimonides's messiah seems to be ordinary in his physicality, even if not in other ways. The emphasis on an exceptional messianic body is chiefly Christian, though David Brakke points out that there are passages in Hebrew scripture that, while not themselves presenting the idea of an incarnate divine messiah, would have influenced that idea's development in Christianity, reminding us again not to mark that particular division too sharply. "The possibility that God or a manifestation of God has the appearance of a human body," Brakke writes, is grounded both in Genesis, where God suggests making a human in their image, and "in Ezekiel's vision of God in 'something that seemed like a human form.'"[4] Christianity brings this God-manifestation explicitly together with the figure of the messiah as one who returns, bringing back to the world something it has lost—some form of unity with the divine, by being the divine in the flesh.

Christianity is almost universally messianic, as one would guess from the similar meanings of the Greek *Christos* and Hebrew *Mashiach*, one who is anointed, that give us *Christ* and *messiah*. Because Christianity is also allegedly monotheistic, the incarnate divine savior poses a puzzle of gathering and division, as creator and redeemer seem both to join and to differ. One result is, as I have mentioned, that Christ may take on a role quite like that of Wisdom, a co-creator who is also read as an aspect of a singular maker and who teaches divinity from within the world. The association of Wisdom with Christ becomes especially strong in Christian Platonism, but it appears already in several passages from canonical Christian scriptures. Paul calls Christ the one "through whom [the Father] created the worlds," who is the "reflection of God's glory and the exact imprint of God's very being, and [who] sustains all things by his powerful word."[5] In Paul's Letter to the Colossians there is a hymn, probably of earlier origin than the rest of the text.[6] "In him all things in heaven and on earth were created," it says of Christ; "things visible and invisible . . . all things have been created through him and for him. He himself is before all things, and in him all things hold together."[7]

We have noted already another passage making very similar claims: the poetic prologue to the Gospel of John: "In the beginning was the Word [*Logos*], and the Word was with God, and the Word was God. He was in the beginning with God. All things came into being through him, and without him not one thing came into being."[8] Augustine, reading the opening of Genesis so clearly echoed in John, will use these references to thoroughly mingle the figures of Christ and Wisdom. We may ask, he says, "whether 'In the beginning' means in the beginning of time in the principle, in the very Wisdom of God. For the Son of God said that he was the principle. When he was asked, 'Who are you?' he said, 'The principle; that is why I am speaking to you.'"[9] God's primal words, Augustine writes in another consideration of Genesis, are "spoken" "by the light of Divine Wisdom, coeternal with Himself and born of Himself." In the same passage, he refers to "the Word, who exists forever in immutable union with the Father," reiterating this point more directly later in the text, where he says that creation was "by His coeternal Word, that is, by the interior and eternal forms of unchangeable Wisdom."[10]

After creation, Christ also takes on Wisdom's advisory role, remaining and circulating in the world to the benefit of its inhabitants. John's text emphasizes that presence in the world: "And the Word was made flesh," it says, "and lived among us," putting the divine into the material and the everyday.[11] Augustine even writes that this embodiment is perhaps the key element distinguishing Christianity from the Platonic philosophy with which it shares so much else.[12] Though John is in many ways distinct from the other canonical Christian gospels, offering a slightly different account of events and putting a far more philosophical spin on its descriptions, it is a key source for the doctrinal idea of a god made flesh. The letters attributed to Paul likewise insist on the fleshiness of Jesus not just generally, but even in his resurrected body, and consequently in the risen bodies of his followers. This resurrection is itself a result of salvation. So in orthodox Christian dogma, the paradoxical jointure of eternal divinity and temporal flesh is necessary for salvation, which mends the bad division between humans and their god. Considerable argument ensues, though, as to *how* the divine and the bodily can work together. Attempts to resolve the apparent paradox, as for example by making the flesh an illusion (albeit a necessary one), are labeled heretical.[13]

These differences mean that there are distinct understandings of the way that the divine body goes about its redemptive work. In the most common Christian narrative, the death of Christ by crucifixion is an unusual version of a sacrificial offering. The first Epistle of John assures its readers that Christ "is the atoning sacrifice for our sins, and not only for ours but also for the sins of the whole world."[14] A sacrifice to a god can be petitionary, but often, as here, it is propitiating. By this remarkable sacrifice of a human-bodied god, the guilt for the original sin of the divided and disobedient human will is expiated. Sacrifice requires a body—a body that can even die (though it will not stay dead). The original humans' sin was so great that only a truly exceptional sacrifice could make up for it; as an act of extreme love, the creator god embodies itself as the sacrificial offering to itself. After the sacrifice, people retain their tendency to act badly, but their efforts to seek and receive forgiveness now have some chance of success (just how much success depends on the particular Christian sect).[15] In most Christian readings, the crucifixion and resurrection do not guarantee salvation, though in chapter 4 we shall see some important exceptions. But they do make it *possible* for the fallen human to return to unity, with flesh rejoined to spirited will and both of those to the will of God. Redemption is accomplished by a combination of sacrificial atonement and the divine ability to move the flesh beyond death into a rebirth or a new life at some later time, in a world yet to come.

Traditions in which the Wisdom influence remains especially strong tend to understand salvation itself a bit differently, in part because of a different understanding of the first problematic division. Here the division is not of human from divine will but of humanity from the particular wisdom that is properly both divine and its own. The primal sin is still a kind of distraction or turning away, but the faithless faculty is less will than attentive memory, the flaw less insubordination than forgetfulness. The function of a savior, therefore, is not to provide restitution but to prompt remembrance. Here Christ is a teacher who helps to turn forgetful people back to the divinity they have forgotten: the Valentinian *Gospel of Truth* says that the very term "Savior" indicates the "work he is to do to redeem those who had not known the Father."[16] This Savior is specifically identified: "The hidden mystery Jesus Christ enlightened those who were in darkness because of forgetfulness."[17] Particularly among the Valentinians, this redemptive possibility still requires the crucifixion, which acts not as a sacrifice but as a "publication," with any number of mythic images coming together at the chiasmic point of a cross.[18] Salvation depends upon an embodied savior, but the redemptive act is pedagogical rather than sacrificial, and it occurs within the world rather than awaiting an afterlife. Christ and Wisdom are two distinct figures in these texts, each taking on some of the roles of biblical Wisdom, working together in the world. The immanence of salvation and the particular positivity that at least some Valentinian texts show toward materiality offer us a particularly rich reading of the redemptive role of Christ as embodied, messianic divinity.

Some aspects of this salvific mystery connect it to much older mystery traditions, and the commonalities can be illuminating. In the Eleusinian mysteries, dating from as early as 1600 BCE, the devotees of Demeter joined with their god. Through their rituals they repeated her trials and shared her divinity. Demeter governs the harvest and the world's fertility, and part of what her followers knew through identifying with her was renewal, the return of life even when the earth seemed dead. Their own divinization and the world's renewal were connected as part of the same enlivening. An echo of this connection can be heard where, as in gnosis-oriented traditions, salvation is memory.

This means that redemption is closely linked to creation, as what is to be remembered pulls back toward a beginning. A divine co-creator present in the world reminds the forgetful of their own divinity. The reminder of the divine is also a rebirth for those who remember. And here too initiates are not transformed, not made new, without the transformation of the world. In the Demeter cults, this means that human divinization is caught up in the cycle of seasons; in the particular form of gnostic memory that I want to explore here, it means that the world must be newly revealed in order for knowers to be new to themselves. The redemptive body is not self-contained. The redemptive knowledge that it brings cannot simply belong to individuals. The rest of this chapter is a slow effort to perceive this revelatory matter of the world. It will be some time before either the materiality of creation or the bodiliness of the saving figure emerges, but the abstract background, though it will demand readerly patience, will enrich our sense of materiality when it finally does appear.

Innumeration

> Knowledge is not made for understanding; it is made for cutting.
> Michel Foucault, "Nietzsche, Genealogy, History"

A new beginning can make no sense without a first. The stories of the first beginning, even just within knowledge-oriented forms of Christianity, are wildly varied and are made more uncertain by the incompleteness of many texts. We find a wide range of narratives even among the Valentinian stories that will be my focus here. I will not at all pretend to present even a single complete account. Instead, I want to look particularly at the places in those stories where limitation plays an essential role, as it does in the bounding and setting apart that characterize the onset of Genesis. As Brakke rightly points out, "When it comes to Gnosis, boundaries can be productive and bodies can be productive means of thinking about boundaries."[19]

It is already apparent that acts of limitation are important across creation stories drawing or building on Genesis. By far the most extended discussion of limit, though, and the most necessary role given to it, come from these Valentinian texts, and from irate descriptions in Irenaeus and Tertullian. Limitation is personified in Valentinian thought, where its productive and creative role emerges in complicated relation to Jesus and Wisdom figures. The nature of the personification is itself elusive. Irenaeus, in his description of Valentinian theology and cosmology, refers to Limit as "the power that went forth from the Son."[20] Sometimes Limit seems to be such a power, sometimes an aspect of the Son or of the Father, and sometimes an entity emanating from one or the other of those. Valentinus and his theological descendants are consummate and poetic myth-makers, so designations are more fluid in kind than they would be in less mythic genres. It is not that one entity changes names or that a name picks out multiple individual beings. It is rather that a name might belong to an entity in some contexts, a power in others, or an act in others still, and this fluidity is not an accident, but part of the poetry of the accounts, a useful reminder that they are not histories or propositions in physics.[21] The theological picture is mobile, less painted than choreographed. This too turns out to be appropriate to the understanding of materiality that finally comes out of these stories of creation and redemption.

The peculiar complexity of Limit and its actions will make more sense if we start with Plato, the ideas' key source. Plato's understanding of Limit has to do specifically with questions of boundaries and divisions, and thus with questions of number. Any Western premodern reference to number is likely to lead into Pythagoreanism, and in fact Plato's work on these questions shows a strong Pythagorean influence. According to Aristotle, the Pythagorean sense of number is that of "a first principle, both as the material of things and as constituting their properties and states."[22] Numbers are not just abstractions; they are the real stuff of the world. But because Pythagoras left no documents and was famously secretive, much of what comes to be called the Pythagorean tradition is really a late ancient recreation of Pythagorean ideas, with considerable influence from a range of philosophical traditions.

Plato is better known not as a lover of arithmetic but as the philosopher who clearly lines up and hierarchizes epistemology along with metaphysics. Perhaps the most famous such line-up is the diagram of the divided line from the *Republic*, where Socrates describes an ascending series of things that exist—from mere images such as shadows all the way through eternal and universal Forms—alongside a matching series of the ways that we know them, from imagination through true intellection.[23] But in his later work Plato, often through Socrates, begins to emphasize something else about knowledge: that it is a matter of collection and division. Collecting and dividing is a method that first organizes, but eventually becomes deeply strange. The corresponding ontology is no less so.

The strategy of gathering and splitting first appears in the *Phaedrus*—in which, uniquely among Plato's dialogues, Socrates ventures outside the boundary wall of his beloved Athens. Here, Socrates declares that we must divide topics "where the natural joints are, and not trying to break any part, after the manner of a bad carver . . . just as the body, which is one, is naturally divisible into two, right and left."[24] It shows up in the working of Eros in Aristophanes's

Symposium speech, where original humans are cut in half by jealous gods and seek to rejoin into wholeness.[25] The image of cutting reappears in the *Statesman*: "For it's all very fine to separate straightaway the thing sought from the others, if it's done correctly.... But really, my friend, it's not safe to make small change: it's safer to proceed by cutting through middles."[26] Both knowledge and love demand accurate division.[27]

Gradually, the process becomes more explicitly mathematical. In the *Theaetetus* and the *Statesman*, Socrates and a visiting student of Parmenides discuss strategies of collecting and dividing with two young students of geometry, who enjoy mathematical challenges.[28] One of them, Theaetetus, looks and thinks like Socrates.[29] The other is named Socrates, and in the *Statesman* the philosopher Socrates remarks on the dual resemblance.[30] It is as if the elder Socrates were divided into two young men, or they collected into him. The dialogues attempt to understand what knowledge is, what good judgment is, what a statesman is—even what those highest abstract Forms are. They worry about the relations of parts to wholes and the One to the Many, and about the inclusion of sets among their own members. Socrates ends the *Theaetetus* by reminding Theaetetus to retain the wisdom of limits, of knowing how much he does not know.[31]

Though the *Theaetetus* and *Statesman* are confusing dialogues, they are not as famously baffling as Plato's *Parmenides*. Philosopher Kenneth Sayre argues for reading the *Parmenides* as the starting point for Plato's shift away from the *Republic*'s hierarchy toward a more arithmetical Pythagorean ontology. In the *Parmenides* as in the *Statesman*, a visiting philosophical Stranger, a follower of Parmenides, talks to a young Socrates. It is not, however, the same Socrates. The young Socrates in the *Parmenides* is the man who will become the philosopher Socrates, and not the geometry student of the *Statesman*. The central question in the *Parmenides* is one dear to Parmenides as a historical figure. It is about Being itself: must Being be singular, or is multiplicity possible? Through a series of increasingly strange arguments, it begins to appear that neither option can work.[32]

A particularly thorny problem is posed by the theory of Forms—the eternal abstractions in which lesser things have their being by "participation." In fact, several of these late dialogues make the theory quite problematic, and it seems clear that this is intentional on Plato's part. We can already see problems brewing in the *Statesman*: "Whenever there's a form of anything, it's necessary that it also be part of the thing of whatever it is said to be a form; but there's no necessity that part be form."[33] The puzzles continue in the *Parmenides*. If one Form is in each of the many things that participate in it, how is it still one?[34] And what makes it the Form that it is, if it does not participate in itself, or is not made to have form in common with the things that participate in it, perhaps by some other form?[35] (The medieval Jewish philosopher Ibn Gabirol will add an intriguing Aristotelian twist to this question by describing not only the form, but the matter, of any object as something in which it participates.)[36] Further problems emerge as Socrates and Parmenides try to consider unity itself. The One, they determine, "will not be a whole, and will not have parts, . . . [and] can have no beginning, or middle, or end." The One is so important because Parmenides in the dialogue affirms a singular first principle, prior even to every form. In fact, for him Being *is* One, rendering the variety of forms beside the point. "Beginning and end," he says, "are the limits of everything . . . the One, if it has neither beginning nor end, is unlimited . . . and therefore without form."[37]

This is not the dialogue's ultimate "answer," however, especially if Plato is indeed developing more Pythagorean ideas here. For the Pythagoreans, there is not a single but a double first source: Limit and Unlimited. Limit and Unlimited are foundational, and each is one singular principle—a sort of double instance of One. Plato crucially complicates this idea even more. As Sayre explains, "Aristotle . . . say[s] at met987b25–27 that Plato agreed with the Pythagoreans save in making his Unlimited two—the Great and (the) Small—instead of a single principle."[38] Aristotle identifies "the Great and the Small" as one of the names for the second half of the pairing more often called the One and the Unlimited (or Indefinite) Dyad.[39] Socrates and Parmenides use the language of Great and Small as they try to work through the puzzles of One and Other than One. From the epistemological consideration of establishing limits—figuring out how to divide in the right places—we have moved into the ontological considerations of a One that delimits and a Two that is unlimited, as co-original principles.[40] Collection and division have led us into a strangely lively and generative sense of number.

Sayre analyzes several textual passages, including those I've mentioned, that hint at what are called Plato's "unwritten" doctrines, where the One and the Unlimited Dyad are foundational. The ideas are called unwritten because, despite hints such as those we have just seen, they do not seem to show up directly in Plato's dialogues—with the possible exception of the *Philebus*, as we shall see.[41] We draw evidence for them both from the hints in those texts and from Plato's successors. Aristotle discusses the concepts in his *Physics* as well as his *Metaphysics*, and so too do Alexander of Aphrodisias and Simplicius in their commentaries on Aristotle.[42] They show up as well in works by some of Plato's own students, including Speusippus, his immediate successor at the Academy, though those texts have been lost.[43] From these sources, we gather that the heart of the "unwritten doctrines" is that the One and the Indefinite

Dyad generate everything (else) that is.[44] The terminology for the primal pair is inconsistent; the One is also called Unity and Limit, and the Dyad is also called Indefinite Two and the Unlimited, and, as we have just seen, Great and Small.[45]

It is not clear that Plato intended these ideas to be secrets. Aristotle's student Aristoxenus recounts learning from his teacher about a "public lecture" in which Plato was expected to explain "The Good." To the bewilderment of the sizable and eager audience, however, Plato spoke instead about mathematics, especially about the relation between one and two—perhaps since, as Lenn Goodman describes it, "for Plato this issue had become the final undissolved residue of philosophy."[46] But it remains uncertain just what kind of event this "public lecture" was. It seems to have been something beyond the usual lectures that the students at the Academy might hear. In fact, it is such an anomalous event that W. K. C. Guthrie declares Aristoxenus's story no more than a fiction.[47] Konrad Gaiser suggests another explanation: perhaps Plato originally intended his more Pythagorean-influenced ideas to remain mysteries, carefully and privately transmitted in proper Pythagorean fashion. When he realized that these ideas had made their way into wider awareness, where they were being passed along full of errors, he gave the lecture as a corrective.[48] Whatever his motivation might have been, and whatever this odd lecture's circumstances were, tradition at least holds that there was a public discussion, and it does seem difficult to describe an idea as simultaneously secret and public. Like a mystery, though, Plato's Pythagoreanism cannot be understood quite straightforwardly, or perhaps without some form of preparation, and the "unwritten doctrines" were taken up in esoteric traditions as if they conveyed a heritage of secrets.

Sayre notes the strange work of delimiting in the *Statesman*, *Theaetetus*, and *Parmenides*. He argues, though, that the strongest case for the idea that Plato's later work is on its way to a more Pythagorean ontology is the discussion of the One and the Unlimited in the *Philebus*. Early in this dialogue, Socrates recommends a method that he attributes to divine wisdom, passed down from ancestral sources.[49] This is a sneaky move—the attribution allows him to circumvent his own insistence that his ideas and abilities are imperfect and uncertain. Socrates claims that he has always been devoted to this method, but, as Sayre points out, it is so unfamiliar to readers that it has generally left them baffled.[50] The method entails some collection and division, and reflection on the one and the many.[51] These aspects will be familiar to Plato's readers. But it focuses most extensively on the Limit and Unlimited (*peros* and *apeiron*—the terminology is Pythagorean). In an intriguing resonance with the elusive public lecture, Socrates offers here the trio of "beauty, truth, and measure," echoing the triad of highest Forms—Beauty, Truth, and the Good—that will be celebrated by neo-Platonism.[52] Measure seems to be what makes good—proper measure is responsible for truth and beauty.

So how do these two principles of Limit and Unlimited, together a double first principle, work? What seems to be most relevant about the Unlimited is not a sense of infinite extent, but an indefiniteness. That is, it is entirely undivided. The One gives the Dyad definition *by limiting it*, making divisions and difference within it. In the *Philebus*, this division introduces proportion to the continuous Unlimited, and this proportionality makes balance and harmony possible. The Unlimited Dyad, though, is not just a sort of muddle. It is dyadic; that is, it is a double in need of division so that its two can be measured or counted. John Turner summarizes the Middle Platonic understanding of this activity: "The One acts by imposing limit on the unlimitedness of the Dyad, . . . When limited by the One, the Dyad, serving as a sort of mold produces the number two, from which the rest of the natural numbers follow by a process of doubling and adding one."[53] To simplify a bit, we might think of an undivided line that is then parsed out into pieces, one by one, the first division making two, or rather revealing the dyadic character of the unlimited. Through addition and multiplication, the one and the dyad, limit and unlimited, generate all other numbers.

We need the indefinite to have something to number; we need limitation to number what was amorphous. That is: it is the One, undifferentiated, that generates difference, within the Dyad, which is two and thus differentiated, but indefinite and thus undifferentiated. Limit can rightly (justly or proportionately) divide the undifferentiated continuity of the Unlimited Dyad. The One cannot be differentiated, as this would create multiplicity, but it must have within it the potency of creating the not-One. The indefinite Dyad cannot be definite, but must have within it the potential, the potency, for definition. Each principle harbors the condition of its own other. Limit is only made Limit when it divides something. Unlimited, as formless, is no-thing without Limits. Within each first source is the whole, which is made of itself and its other, and which is not fixed as a totality but is constantly in production. And though language's limitations make it simplest to call this a pair, it remains One *and* Two. From the division and combination of these, the world emerges.

Plato has Socrates describe division most extensively, but collection does not disappear, and it is no wonder collection and division are puzzling. When we collect or gather together, we are performing two actions: drawing into one and excluding all others. But pure Oneness cannot exclude any others, since exclusion always makes two. When we divide, we distinguish one from another. But the principle of division, or limit, belongs only to the One; it can never be included within the categories that it makes, but neither can it contradict them. The Unlimited Dyad could contain many, but they cannot be differentiated except by Limit, so their multiplicity is without difference. In other words, the confusion induced by the later dialogues might be a hint to us that when we collect and divide, and when we then think these acts through to their ontological foundations in Limit and Unlimited, we really do generate and encounter paradox, but this paradox is not mere nonsense.

As the Platonic tradition develops, the Pythagorean element intensifies, and both mystery and number become even more important. In *The Theology of Arithmetic*, Joel Kalvesmaki describes two understandings of number set forth in the late first century CE by the neo-Pythagorean Platonist Moderatus of Gades. The first is a theory of stable sets or collections, "for example, a constellation of stars, a pile of pebbles, or some other inert collection," in which number measures the multitude, and each item in the collection is a distinct individual thing.[54] The other, which Moderatus associates with Pythagoras, is livelier, with number actually generating multitudes, not just measuring them. In this sense one is not just distinguished from the one next to it, but added to itself to make two, or added to a pair to make three, and so on. The first sense allows us to count out a stable manyness as a series of one after another one, trusting that each unit will stay where it is, and even that counting is an act that could be finished. The second sense is "suggestive of movement, transition, change, and cycles. It invokes metaphors of procreation and generation."[55] This understanding allows the neo-Pythagoreans to see number as generative of the cosmos. As Kalvesmaki points out, "Plato embraces this definition when he explains how the one and the indefinite dyad generate the numbers."[56]

And for Pythagorean and neo-Pythagorean thought, numbers generate everything. The slightly more familiar idea of the originary One will be central in neo-Platonism, where its creative emanation can best be explained as the natural tendency of what is perfect to exceed itself. Plotinus, the exemplary neo-Platonist, writes, "Seeking nothing, possessing nothing, lacking nothing, the One is perfect and, in our metaphor, has overflowed, and its exuberance has produced the new."[57] From this perfect One, limiting is productive for Plotinus; it even enables unity, the state of oneness that is as close as other things can come to that first perfection. "Life looks toward the One and, determined by it, takes on boundary, limit and form," he writes. "Each element of multiplicity is determined multiplicity because of Life, but is also a Unity because of limit."[58] Unity, whether original or regained, is understood to be the best possible state. It generates because it is in its nature to do so—perfection naturally exceeds itself—and all multiplicity redemptively returns to it. The unity of a thing is given by limit.

The pairing of One and Indefinite Dyad that we find in some Middle Platonism is not entirely different, but neither is it quite the same—and that neither-sameness-nor-difference echoes the essential, constant, and generative ambiguity of the pair itself. The idea of an original One and Unlimited Dyad is taken up in different ways by different groups. In some versions, the Monad produces a Dyad; in others, they are absolutely co-original.[59] Turner writes, "According to Hippolytus, both the Valentinians (*Ref.* VI.29.5–6) and the 'Simonian' *Megale Apophasis* (*Ref.* VI.18–4–7) used the concept of the emanation of a dyad preexisting in the monad. . . . In the *Apocryphon of John*, Barbelo is derived from the Monad as the product of the former's self-reflection."[60] Co-originality appears again in medieval Kabbalah. In Elliot Wolfson's exegesis, it goes beyond "the standard Neoplatonic emanationist scheme that begins with the absolute and indivisible One from which proceeds a second being that is itself one but also two, a one that is many." Instead, "the oneness of God is a unity in multiplicity."[61]

John Dillon traces our limited knowledge of the historical development of some of these readings. Plato's successor Speusippus understood the One and Dyad as connected to levels of reality. In his thought, "the same entity manifests itself on successive levels of reality. For Speusippus, Multiplicity (*plethos*) already appears at the level of the One 'as a completely fluid and pliable matter.' It continues to manifest itself at each successive level of reality: arithmetical, geometrical, psychical, physical—until it appears as matter proper."[62] We can recognize in this perfect fluidity the unlimited interactive character of matter in the *Secret Book of John*. The Dyad is often associated with materiality. Of the Platonism of Moderatus, Turner writes, "In order to generate determinate being, this indefinite material or Quantity must be limited by form, but the unitary Logos first has to deprive itself of all traces of its unitariness in order to admit or make room for Quantity, in which act it becomes formless and shapeless itself."[63] Moderatus's speculation about matter, Turner notes, is in turn "fundamental to the understanding . . . of Sethian and Valentinian Gnosticism" over the next few centuries.[64] Continuing the passage cited earlier, Wolfson describes a strikingly similar self-limitedness in Kabbalah: "There must be an aspect of the Infinite that transcends even the attribution of unity since *to be one, the one must be two*."[65] As these few examples suggest, the Pythagorean influence emerges as myth with all the precision of mathematics.

The Dyad's association with materiality is not always positive. The Pythagoreans, who loved binaries, seem to have connected the unlimited to such dangerous attributes as darkness and femininity.[66] Femininity may even cast doubt on the workings of Wisdom. Turner argues that Philo, for instance, would greatly have preferred to make Wisdom, who is grammatically feminine

(*Sophia*), into a proper masculine educator, but he could not "completely escape the influence of the Jewish tradition of the femininity of Wisdom nor the Pythagorean tradition of the femininity of the Dyad."[67] In most gnostic traditions, Wisdom will be distinct from the Dyad, and more distinctly gendered, but both will be mobile and generative to a degree that unsettles numerical counting together with neat cosmic order, and neither will allow matter a simply profane status.

Kalvesmaki points out that the mythic use of Pythagorean ideas is not an attempt at strict fidelity. For the Valentinians in particular, "Pythagoreanism was a literary tissue, not a community, an intellectual skin that the Valentinians stretched and altered as much as they depended upon it."[68] One consequence of this literary approach is an apparent inconsistency in the choice of monadic or dyadic theologies and cosmogonies. Hippolytus writes that there were both monist and dualist Valentinians and that they were in significant disagreement. Unhelpfully, he focuses his criticisms exclusively on the monists, though a sharp dualism of good and evil is a more common criticism of gnostic groups.[69] Irenaeus takes a more nuanced view, if hardly a friendlier one: perhaps the primal Depth at the origin of Valentinian cosmology is wholly solitary, he says. Perhaps, though alone, it is double in being hermaphroditic. Or perhaps it is paired—with one consort or with two.[70] For Irenaeus this variability shows the weakness of Valentinian Christianity: it cannot even be consistent with itself. Tertullian similarly accuses the Valentinians of vacillating between monism and dualism, among other contemptible conceptual moves.[71]

Valentinian writers tend not to be much concerned with choosing consistently between monadic and dyadic motifs. Besides the literary license that the texts clearly take, there are variations that develop over time and differences between Eastern and Western versions. But the stories might also vary because it is in fact the One and Dyad *together* that the writers find most generative—and to be either simply monadic or simply dyadic loses part of the paradox's productivity. To develop this suggestion, I will be drawing most extensively from Irenaeus's detailed descriptions and from the *Valentinian Exposition*, a possibly catechetical text, where the paradoxicality seems to me most evident. But I will make use of other texts, too, to point out some of the mysteriousness that emerges in matter across Valentinian sources.[72]

For instance, the Valentinian *Tripartite Tractate* declares, "The Father is a single one, like a number, for he is the first one and the one who is only himself. Yet he is not like a solitary individual. Otherwise, how could he be a father?"[73] The *Valentinian Exposition* tells us, "[They] saw him residing in Oneness, in Twoness, and in Four, bringing forth the Only One and the [Boundary]."[74] And even Western orthodoxy, in the writings of Augustine, retains some of this remarkable sense. Continuing the passage in which he identifies the Son with Wisdom as the first principle, Augustine writes, "For there is the principle without principle, and there is the principle along with another principle. The principle without principle is the Father alone, and thus we believe that all things are from one principle. But the Son is a principle in such a way that he is from the Father."[75] Principles are foundational, and the foundation is at once singular and double—and, perhaps, at once co-originary and emergent. The condition of its other is within each principle, as we have seen, and that double condition is necessary for the possibility of creation.[76]

What is simply unlimited is unknowable. It is limit itself that tells us of the unknowable unlimited. To know limit, though, is itself more difficult than it seems. In Valentinian cosmogony, the world comes about in a series of limitations—some generative, some temporarily destructive, some redemptive. In this series, as I have suggested, the power of limit entangles itself with the body of the savior. We are almost ready to see how. But because creation and salvation are one long process in Valentinian cosmology, we will have to go through several stages of creation by Limit in order to come at last to redemptive flesh.

The Triple Work of Limit

First Delimitation: The Pleroma

> She wraps herself in boundlessness and exits.
>
> Anne Carson, "Contempts"

As we can already see from reading the heresiologists, it is impossible to find a singular Valentinian origin story. The *Gospel of Truth* repeatedly places at the beginning the "illimitable, the inconceivable one."[77] But any apparent unity in the description is very strange: "All have been within him, the illimitable, the inconceivable, who is beyond all thought."[78] If this One has others within it, how is it One? If the All are inside the One, how do they ever become anything that is not the One? If the One is illimitable, how is anything, including the All, ever to be divided from it?

The *Valentinian Exposition* is probably later than the *Gospel of Truth*, and its cosmogony is more complex and more clearly neo-Pythagorean. It speaks of the beginning as "the Father, who [is the root] of the All and the [Ineffable One]." This One "exists as Oneness, [being alone] in silence—'silence' means tranquility—since [he was] in fact One, and nothing existed before him."[79] This seems to establish an original singularity. But then we read, "He also exits [as] Twoness and as a pair—his partner is silence."[80] One is both alone in and partnered by Twoness that is silence. So one is two without loss of unity; two is one without loss of pairing. And then, as in the *Gospel of Truth*, we read, "He possessed the All dwelling [inside] him."[81] How many is all that is, dwelling in one and in two?

So now there is a oneness that is a twoness, a solitude that is paired, a within that can have no without. Though they will not place the same emphasis on the role of Limit, other kinds of Gnostic texts present us with further versions of the puzzle of this firstness. *Zostrianos* says that the real God "remains alone in himself and rests himself on his limitless limit."[82] In the revelation described in *Allogenes*, One and Unlimited are together: "And I saw an eternal, intellectual, undivided motion that pertains to all the formless powers, [which is] unlimited by limitation."[83] The *Secret Book of John* says, "The One is without boundaries. Nothing exists outside of it to border it. . . . The One cannot be measured. Nothing exists external to it to measure it."[84] What limits other things cannot itself be bounded. Kalvesmaki groups this last text among those of "the Barbelo-Gnostics," the name that Irenaeus gives to the Sethians. He describes them as "a group related to the Valentinians," though, as he points out, the exact nature of that relation is not quite clear.[85] He continues, "According to Irenaeus and Theodoret, they held that the uppermost aeon, the unnameable Father, dwelt within the second one, the virginal spirit called Barbelo. But neither apologist further explains the relationship between the two uppermost aeons."[86] (Though it is not the most likely, one proposed etymology of *Barbelo* links it directly to the Valentinian *Horos*, Limit, as "supreme limit.")[87] As Kalvesmaki notes, the *Secret Book of John* thus presents its reader with a monad and a dyad that are neither a monad nor a dyad in any comprehensible way.[88]

We know from Plato that the One must somehow limit the unlimitedness of Two. From the Monad and the Pair that are One and Two, says the *Valentinian Exposition*, first "God [came] forth, the Son, Mind of the All."[89] In this text the Son is "the revelation [from the] silence, and a Mind of the All, being Two with Depth."[90] This Mind precedes the All. The All is the Aeons in their realm. The subsequent movements of the Son are summarized quickly before they are later elaborated: "For he is the producer of the All and the actualization [of the thought] of the Father—which is [Desire]—and the descent down below. When the First Father willed it, he revealed himself in him."[91] That is, the Son will be the aspect of divinity that reveals the Father in the "lower" world with which we are familiar. Christian readers would easily recognize the salvific function of Christ.

In this same passage, Limit (translated here as Boundary) emerges, brought forth by Truth.[92] Only through Limit can the All come to be. "And the Boundary . . . [separated] the All . . . is totally ineffable to the All, and the confirmation and actualization of the All."[93] It is through the Limit that All is made (by being made other than the Father). That is, Limit is what makes the All other than the Father. This text also tells us that Limit has the powers of separating, strengthening, giving form, and making substance—the powers of making and sustaining things.[94] Though it is easy to get lost in these details, this point is also rather commonsensical: a thing is what it is because it has particular limits, where it is not merged into something else.

Limit is necessary for the All, but is, like the Father, beyond the All's understanding, "totally ineffable to the All." In fact, the Limit is "[the one] who contains the All, the one who is elevated [above the] All."[95] This suggests that Limit resembles the Father and the Son as being beyond what exists. Limit is not just All-making, but All-encompassing, and therefore beyond comprehension. The Limit is further described as "the veil of [silence]. He is the [true] High Priest, the [one who has] the authority to enter the Holies of Holies."[96] Limit divides like a veil does, as a concealment uniquely able to perceive both sides of itself, much as the Son can perceive both the All and the Father. Limit keeps the Aeons of the All distinct from the source, allowing them their existence; it keeps their knowledge of the origin incomplete, allowing them their desire. Thinking and desiring work together if Desire is the first thought of the Father. Limit teaches, "revealing the glory to the Aeons and bringing forth the abundant wealth."[97] Revelation is redemptive. The Son reveals the Father in the world. As the power coming forth from the Son, Limit reveals the Father's glory. Through the Son's delimiting power, all things are made. And they can only be made as they are—other than other things—by Limit. Brakke points out another paradox created by Limit's dividing processes. On the one hand, differentiation generates distance and withholding, resulting in ignorance and unfulfillment. On the other, existence itself and "the distance that genuine knowing and loving require" are generated by that same division.[98] As Anne Carson has suggested, desire makes and depends upon edges.[99]

The role of Limit will be emphasized again later in the cosmogony. So far we have a first delimitation, allowing the Aeons, who are the members of the All, to be something other than the Father. Limit acts like the One in Plato's *Philebus*, making form by dividing an otherwise unlimited that is a potential multiplicity. And we have a hint that the All and the origin will be, however indirectly, revealed through Limit. Thus it is that the Limit, which keeps things apart, is the image of the One, which cannot have parts, and of the *Unlimited Dyad*. The Son is the first from and always with the Father. There is nothing between Limit, as a power or aspect of the Son, and Father (what is between could only be another limit). And Limit is between Father and everything else, keeping the

All separate within. Interpreting the canonical Gospel of John, Theodotus writes, "When [Jesus] says, 'I am the door,' he means that 'you . . . will come up to the boundary where I am.'"[100] What is at the limit is the Limit itself.

So the first work of Limit, who is a power or aspect of the Son who is with the Father, is to create the All or the Pleroma, the realm of the Aeons. The second will be to divide the Aeons themselves.

Second Delimitation: Wisdom's Divisions and the Making of Matter

> And all of us, and everything
> Already there
> But unconstrained by form.
>
> Rebecca Elson, A Responsibility to Awe

The limitation of the Aeons' knowledge of the Father preserves them, because unlimited knowledge would make them one with him, and they would lose their own being. The problem is that, like most of us, at least one of the Aeons *wants* to know more completely. There is nothing wrong with this desire. In fact, the *Tripartite Tractate* elaborates on the claim that the creation of the Aeons is designed to sustain it: The unfathomable one "did not wish that they should know him, since he grants that he be conceived in such a way as to be sought for, while keeping to himself his unsearchable being."[101] Irenaeus writes that in Valentinian cosmology the origin wills "to create within [the Aeons] a desire of investigating his nature . . . a wish to behold the Author of their being, and to contemplate that First Cause which had no beginning."[102] Tertullian's language is stronger still; the Father "wanted everyone to be inflamed with a desire for him," such that the Aeons "burn with silent longing to know the Father."[103]

In the case of Wisdom, this wish becomes dangerous. Irenaeus writes, "On account of the vast profundity as well as the unsearchable nature of the Father, and on account of the love she bore him, she was ever stretching herself forward; there was danger lest she should at last have been absorbed by his sweetness, and resolved into his absolute essence."[104] Both erotic and intellectual desire create this danger. As Tertullian continues the story, the absorption nearly happened: "Indeed, she barely missed being swallowed up in the great sweetness of the Father and the labor of her search and she almost was dissolved in the primal substance."[105] Wisdom loves the unknowable source so much, and is so curious about it, that she wants to meet with it, to know it—with disregard for Limit, by which she has her very being. And she might have "succeeded" in her own destruction, "unless," as Irenaeus continues, "she had met with that Power which supports all things, and preserves them outside of the unspeakable greatness."[106] This power is Limit. Tertullian elaborates: "Only her destruction would have stopped her search if she had not by a stroke of luck run into Horos [Limit]. (He acts as support and as the boundary-guard of this whole world.) . . . He causes Sophia to become calm by removing her from danger and by persuading her to stop her useless search for the Father."[107] In Tertullian's version, Sophia then divides herself, abandoning her dangerous inclination and the part of her that includes it. In most other versions, this division is performed by Limit. Limit excludes Wisdom from the realm of the All, and she and her consort are separated.

Though it seems punitive, this delimitation is (also) an act of love.[108] Wisdom sought carelessly, risking without knowing it her own annihilation. The source loves with care, where love cannot be the same as completeness or resolution. Wisdom's very existence is preserved when she is set apart from the rest of the Aeons. Limit "restrains and supports" Wisdom until, as Irenaeus puts it, she can be "brought back to herself."[109] But preserved or not, she is lonely and distressed. She weeps, and the other Aeons join in her pleas for help.[110] Limit comes to the rescue once more, dividing Wisdom herself into higher and lower parts. Higher Wisdom is reincluded in the All and returns to her consort.[111] Irenaeus offers a fairly detailed account of this part of the process. In his version, the dividing and forming agent is now described as that of which it is a power or aspect: as Christ, the Son. "But the Christ dwelling on high took pity upon her; and . . . he imparted a figure to her. . . . Having effected this, he withdrew his influence, and returned, leaving Achamoth [lower Wisdom] to herself, in order that she, becoming sensible of her suffering as being severed from the Pleroma, might be influenced by the desire of better things."[112] This figure-imparting is recognizably a function of Limit, the separator and form-provider.

The exiled Wisdom, because "she could not pass by Limit," resigns herself to experiencing a range of passions.[113] She grieves at her unmet desire, fears for her survival, is bewildered by her ignorance, and longs to return to her source—which she remembers with an occasional glimmer of joy. This last, with its traces of memory (and, in some versions of the story, repentance), gives rise to soul, while her other emotions create matter.[114] In the *Gospel of Truth*, "terror grew dense like a fog" in the presence of ignorance of the Father, making matter a consequence of emotion even within its somewhat simpler cosmogony.[115] Tertullian describes a poetic variation. Here different instantiations of materiality develop from Wisdom's various passions—liquids from her tears, solid materials from her fear and alarm, and delightfully (though Tertullian is clearly resistant to this delight), sunlight from her rare smiles.[116]

There are other versions of the story, in which the lower Wisdom makes matter not from emotion but as a still-defiant attempt to create on her own. But on each telling, the matter she makes is not yet any things, because it has not been delimited; it is amorphous. So the first act of Limitation is to set apart the Aeons from the Father. The second is to cut Wisdom off from the other Aeons, for her own preservation. The continuation of this act divides Wisdom into two parts, one still kept away from the realm of the All and the other included back within it. Excluded Wisdom makes matter. In one more step, matter will receive delimiting form.

Third Delimitation: Giving Form to Matter

> You have conferred multiplicity,
> you have found and remained One,
> while still conferring multiplicity through division.
> "The Three Steles of Seth"

Formless matter is no mitigation of loneliness. The stuff that Wisdom has created is still chaotic, in need of formation. Before matter can be divided, though, the Savior must be collected.

The Savior is not a particular Aeon, but gathers all of their aspects together and is sent forth both by the father and by all of the Aeons. Hippolytus calls him the "Joint Fruit of the Pleroma." In him, all the powers of the Aeons are gathered into one, or, from the converse perspective, he is the one of whom all the Aeons are aspects or parts.[117] Like the relation of Son and Limit, this description is slippery. J. P. Kenny remarks with some understatement, "The relationship of Son to the Aeons appears to have a wholepart character, although the exact nature of this conjunction is not entirely clear."[118] The Savior appears to be the Son insofar as it descends to assist Wisdom, but under a subtly different description. The Son was always undivided, while the Savior is a sort of collective of what had been many. The Son/Savior has the power of Limit. Discussing the *Tripartite Tractate*, Brakke points out the complications of even trying to identify an original of which the Savior is a likeness:

> The Savior emerges as an image of the Father, and thus as his Son, but seems also to embody the fullness. The Savior is generated from the harmony and joy of the aeons as their "fruit," but he reveals "the countenance of the Father"; thus he is "the revelation of his (the Father's) union with them (the aeons), that is, the beloved Son. . . . The Savior was a bodily image of only one thing, namely the Entirety. . . . He . . .

also proves to be a bodily image of the spiritual realm, through which the aeons know the Father.[119]

The Son, the first to flow forth from the Father, is now present as the new image of the Savior, bearing the aspects of every Aeon—except for that missing part of Wisdom—and performing an important function of the Son and/as Limit, which is the revelation of the Father. It is more useful not to try to demarcate Son, Limit, and Savior as a set of entities, but to think of the terms as different names meant to emphasize different aspects and functions of what is next to the Father, the origin that is the mutual indwelling of One and Two.

The Savior goes now to aid lower Wisdom. (Though the stories do not say so, perhaps he, as the gathering of the Aeons, feels her absence, missing her as she misses the All.) The *Valentinian Exposition* says of the Savior Christ, "And he possesses four powers: a separator and a confirmor, a form-provider and a substance-producer." These are recognizable powers—we saw them, in an earlier passage, as the powers of limitation. "But why a [separating] power, a strengthening one, a substance-producing one, and a form-giving one, as others have [objected]? For [they] maintain that the Boundary has only two powers, a separating one and a strengthening one, insofar as it separates [the Depth] from the [Aeons]."[120] This distinction, which seems to set Limit and Savior apart, may emerge from a desire to clarify just which functions of thing-making are limitations strictly speaking. It may also emphasize their relation to the cross, which will be relevant shortly. But it is unquestionable that all four powers work creatively together.

Irenaeus specifies that the Savior comes forth from the All, endowed with power by the Father, so that "by him were all things, visible and invisible, created."[121] This is odd, since it seems that matter—visible things—is already created by the passions of Wisdom. But as we have noted, before the action of the Savior, matter is not yet things.

Matter is not left in chaos, nor Wisdom in sorrow. The descended Savior and Wisdom interact generatively. They engage both unformed matter and the amorphous seeds of the spirit that Wisdom has scattered in it. As the *Valentinian Exposition* explains, "Now, since the seeds [of] Sophia were unfinished and without form, Jesus conceived a creation of [this] kind: he created it from the seeds, with Sophia working together with him."[122] We get important clues as to the forms that they make: "Since they were . . . without [form], he came down [and revealed to them] the Fullness. [He instructed] them about the place [of the] uncreated. All things [he made after] the type of the Fullness and the Father, the Uncontainable One. . . . The creation . . . is the shadow of the preexistent."[123] The seeds belong to spirit, but matter is likewise formed after a preexistent pattern: "Jesus, then, fashioned the creation and performed his work from the [material] passions surrounding the seeds."[124] Creation follows the model of the Fullness, the All, which is formed after the pattern of the uncreated source embodied by the Limiting Savior. That pattern, we recall, is the strange pair of One and Unlimited Dyad.

The account in the *Exposition* has Wisdom's better passions introduced into spiritual, and the worse into carnal, things. But even here the shaping and copying are pervasive. The Savior and Wisdom give matter forms "in order to put forth shadows and images of the preexistent, the present, and the future things, that was the economy that [was] entrusted to Jesus."[125] The text already hints that there is some redemptive power even in the images and shadows. In the *Tripartite Tractate*, these "manifest images of the living visages" are described as "resembling them in beauty, but unequal to them in truth."[126] They are not the equal of the uncreated realm, but they are not altogether unlike that realm, either. Believing in Jesus, the *Valentinian Exposition* explains, helps those believers to notice the shadows of the uncreated, though we still need to discuss how this might happen.[127]

Working together with the material that Wisdom has created and the Savior's power that is Limit, Wisdom and the Savior make the created world into images of the All and even provide hints of the absolutely uncreated origin, by separating chaotic matter and providing it with imitative form. Because of this, the seeds of the spirit they have scattered within matter will have a nudge toward remembrance of the highest truth. That is, those seeds are within material beings, not in some alternate realm, and are surrounded by matter formed into the image of the divine fullness, the All, "as through a mirror image."[128] Such an image is inferior to its original source, but it remains an image of beauty, and it presents that source in such a way that more beings can enjoy it. We can emphasize either the inferiority of a reflection to its original or its value as presenting that original to more viewers; it is at least possible, some scholars argue, that Valentinus himself held the latter view.[129] Certainly it is a view that offers us productive possibilities.

This is where we return more explicitly to the One and the Dyad, with matter at the end of creation. The created, material world is the image of the All. The All are the collective aspects of the Limiting Son who is the image of the Father, who is partless one and unlimited two. As a generative restriction—a maker of manyness—Limit creates the All of the Aeons, the realm of Lower Wisdom, and the shapes and beauties of matter, each in its way an image of the singular-and-dyadic source. Without the fluid multiplicity of the Unlimited Dyad, the One would remain only One, and even the All could not be delimited from it.

The material world is, rather literally, a form of divinity. That is, it is shaped in the form of the divine. Matter's quality of being an image of the All is manifest in its beauty. Most of matter may spring from suffering, but it is shaped in the image of joy and lit by the delight of Wisdom. We hear a more joyous echo of the rabbinic commentaries on the built-in nature of forgiveness for sin—the world is made such that it aids the memory of the forgetful.

The connection of creation to both Limit and Wisdom is not restricted to Valentinian or gnostic sources. In Origen's cosmogony, God first generates a certain number of immaterial beings—a number that must be countable, because only God, the illimitable and ineffable source, can be without limit. Once more matter comes about in a second limitation: "*God* has arranged all things in number and measure; and therefore number will be correctly applied to rational creatures or understandings. . . . But measure will be appropriately applied to a material body; and this measure, we are to believe, was created by God such as He knew would be sufficient for the adorning of the world."[130] Matter is made to measure as a reminder. Beauty is divine, and the beauty of material things adorns the world. Limitation can easily sound like privation, but in these stories it offers existence as a gift from a divinity that is not a thing itself. And it offers beauty, because the limiting shape is an imitation of the divine source.

Creation proceeds through the power of Limit, in the interplay of many and one. It divides the All from the Father, Wisdom from the risk of disappearance, and matter from formlessness. The result is imperfect, but it is also beautiful. Moving in this beauty, dwelling among us, we will finally find a savior that is embodied.

Salvific Teaching and Somatic Text

> Jesus said, "On the day when you were one you became two. But when you become two, what will you do?"
>
> *The Gospel of Thomas*, logion 11

Matter provides a place for instruction, where the seeds of the spirit that Wisdom has generated can take the necessary time to remember their divine source—just as Wisdom herself had to do.[131] But perhaps more importantly still, matter as the image of the All and the uncreated calls, through its beauty, to the memories of creatures, calls with the voice of the divine. It gives not only the time for remembering, but a reminder itself. This is almost enough, but the voice of beauty is a subtle one. Redemption requires one more step. Listeners must learn how to hear that voice, how to pick up on the reminders.

This is where (at last!) the *embodiment* of the Limiting Savior who is the Son becomes especially important.

The Savior has descended to aid Wisdom, the two of them working together to make the world a mnemonic aid. In the next step, he descends again to help those living in that world out of their forgetfulness. The matter of the world is one reminder; the material body of the savior is another. The second helps us to interpret the first. Though Valentinian doctrine joins the carnal Jesus to the savior only at baptism and not at or before birth, it does not deny that the redeemer takes on the body. And, like proto-orthodox forms of Christianity, it even understands the church as an extended form of that same body. The claim that the members of the church can also be understood as the members of an extended body of Christ has a scriptural ground in Paul's letters, and Augustine particularly develops the metaphor.[132] In Valentinian Christology, as David Brons explains, "the incarnation is . . . not merely the joining of Christ to Jesus, it is in effect the simultaneous redemption of all who are part of his body."[133] Theodotus says that "the body of Jesus . . . is of the same substance as the Church."[134] Each member of the church is a member of the divine body.

This whole-part relation is once more complex. It is not obvious how many of those members there might be. Gnosticizing forms of religion are sometimes criticized, and sometimes rightly, for a particular kind of elitism that approaches predestination. Those with the right sort of soul are capable of remembrance and thus of redemption, and this can mean that most people are irredeemable. Valentinian texts do suggest that some attain gnosis more readily than others. But Brons draws on several Valentinian works to point out that this redemptive knowledge is of no less value when attained with more effort:

> Valentinian teachers frequently caution those with gnosis to "share it without hesitation" (*Interpretation of Knowledge* 15:36). They are not to despise others as inferior or ignorant, for "you are ignorant when you hate them and are jealous of them, since you will not receive the grace that dwells within them, being unwilling to reconcile them to the bounty of the head" (*Interpretation of Knowledge* 17:27–31). Rather, as "illuminators in the midst of mortal men" (Letter of Peter to Philip 137:8–9), they have a duty to aid in the salvation of those who do not yet have gnosis. In the *Gospel of Philip* it says, "Whoever becomes free through acquaintance (gnosis) is a slave on account of love towards those who have not yet taken up the freedom of acquaintance (gnosis)" (*Gospel of Philip* 77:26–29).[135]

On this model, redemptive memory is decidedly *not* restricted to a chosen few. Those who were not previously familiar with truth may come to believe the true gnosis "because of human testimony."[136] Thus Valentinian teachers can bring redemptive knowledge to anyone willing to listen and learn. In fact, Tertullian is outraged by their willingness to do so: "To begin with, it is doubtful who is a catechumen, and who a believer; they have all access alike, they hear alike, they pray alike—even heathens, if any such happen to come among them.... All are puffed up, all offer you knowledge." He seems to be especially scandalized that even women can be teachers: "The very women of these heretics, how wanton they are! For they are bold enough to teach, to dispute, to enact exorcisms, to undertake cures—it may be even to baptize."[137]

The model of a Christian teacher is, of course, Christ, of whom the *Gospel of Truth* tells us, "He came forward and spoke the word as a teacher."[138] This text offers the most extensive account of the act of teaching itself, and that act is an embodied one. "Acquaintance from the father and the appearance of his son gave them a means to comprehend," the text continues, describing the responses of the taught. "For when they saw and heard him, he let them taste him and smell him and touch the beloved son."[139] The emphasis on the senses makes it impossible to believe that redemptive knowledge can only be an abstraction. By being in a body and bringing memory through the senses, the savior saves bodily beings. Sense memory is awakened along with more abstract recollection. Members of the son's body—that is, of the church—are called upon to bring memory to others too. Bodily acquaintance is a path to understanding. The encounter with the embodied Christ is a sensory connection with divinity—"He let them taste and smell him." With their senses awakened and alerted to the divine, perhaps those granted this privileged acquaintance can go on to spot divinity in other matter as well, noticing that it too is beautiful.

But, the *Gospel* goes on to say, not everyone is impressed by this teaching. "Those wise in their own eyes came to test him, but he refuted them, for they were foolish, and they hated him because they were not really wise."[140] In these lines we can hear not only the more canonical scriptural accounts of Jesus, but an echo of Plato—this teacher sounds very like Socrates, who famously questioned wise people and determined that they were never as wise as they thought they were, since they lacked the knowledge of limits and of the infinite elusiveness of divine truth.[141] The truly wise know both their limitations and the workings of Limit, and in knowing this, they know what they can of divinity. Brakke points out the similar function of the divine bodies in the *Untitled Treatise*, which "may seem to reveal the divine to its reader, but... do so only by marking the limit or boundary of human knowledge. Here boundaries may create, mark, or reveal ignorance, but they also form the only condition for knowing what cannot be known."[142] For human wisdom as for Wisdom itself, this limitation sustains seeking and desire, avoiding stasis. We recognize echoes of both the role of Limit in the All, where it is at once the condition and the restriction of knowledge of the divine source, and the role of the Son who has the power of Limit, who comes into the world to reveal the Father.

We shift here to what looks like more abstraction, but it will soon return us to matter. To those who can listen, the embodied savior teaches reading. The book, however, is an unconventional one. "In their hearts," says the *Gospel of Truth*, "the living book of the living was revealed, the book that was written in the Father's thought and mind and was, since the foundation of the All, in his incomprehensible nature."[143] In the book, the readers' own names are written, and were written before those they named existed.[144] They learn to read what cannot be comprehended, and they learn to read it within themselves, as their own names, "in their hearts." Those whose opinion of their own human knowledge is too high will be shown up as foolish by this incomprehensible truth. The more humble students will learn to know themselves, and thereby to perceive divinity: like Adam in the *Secret Book of John*, they can recognize divinity "without" when they have known it "within." The "living who are inscribed in the book of the living," says the *Gospel of Truth*, "learn about themselves, receiving instruction from the Father, returning to him."[145] They learn this by having the book itself teach them to read that they are also divine.

This is because the book is not one of ordinary pages. So that it can be revealed, the book is bodily. Like matter in the image of the All, this book in which names are written is a reminder, and its instruction comes to us through our senses. Jesus, who gives people a sensory encounter with the divine, also allows them to read himself. "No one had been able to take up this book, since it was ordained that the one who would take it up would be slain.... Jesus appeared, put on that book, was nailed to a tree, and published the Father's edict on the cross."[146] An incarnate, materialized Son, wrapped in the skin-document of the living book, presents divinity so that material human beings can grasp it. "Reading," the means of knowledge and the awareness that one's own name is written and called, is a way of seeing, hearing, touching—of sensing and interpreting, of responding to the call formed in the material world. The book can only be published in flesh, and it would otherwise go unread.

The layering of levels here is extraordinary. The body on the cross is the published book; the book is the many-membered body of the church, containing all the members' names; and the son, whose (many-membered) body it is, is the image of the Father, stretched out bodily both to mark a boundary and to draw back together. This double stretch is a function of the cross, which is not merely an instrument of death. And those written in the book learn from it how to read themselves and the world in which they are already implicated.

The body becomes a book, and the book a body: the word is made flesh, and it is published on a cross.[147]

The site of publication is crucial. The cross itself performs multiple functions, and is listed by both Irenaeus and Tertullian as one of the names for Limit. Irenaeus says,

> Horos [Limit] . . . , whom they call by a variety of names, has two faculties,—the one of supporting, and the other of separating; and insofar as he supports and sustains, he is Stauros [Cross], while insofar as he divides and separates, he is Horos. They then represent the Saviour as having indicated this twofold faculty: first, the sustaining power, when He said, "Whosoever doth not bear his cross (Stauros), and follow after me, cannot be my disciple"; and again, "Taking up the cross follow me"; but the separating power when He said, "I came not to send peace, but a sword."[148]

Tertullian too tells us that Limit goes by other names: "They also call him Cross, Lytrotes (Redeemer) or Carpistes (Emancipator)."[149] In a somewhat complicated passage, Tertullian draws together the division of Wisdom, Limit as Cross, and the bodily Christ into which the Savior has entered: "The soul-like and bodily Christ suffered to illustrate the experience of the higher Christ who was stretched on Cross, otherwise known as Horos, when he shaped Achamoth [Lower Wisdom]." Tertullian has no patience for such rampant mythologizing: "In such a way everything becomes an illustration or image; even, obviously, these Christians themselves are imaginary."[150]

Theodotus complicates the metaphor still further, conceiving the horizontal crossbar as the burden-bearing shoulders of the Savior, whose body is above as well as below that bar: "The Cross is a sign of the Limit. . . . Jesus by that sign carries the Seed [that is, those who have the seeds of the spirit distributed by Wisdom] on his shoulders and leads them into the Pleroma. For Jesus is called the shoulders of the seed and Christ is the head. . . . Therefore he took the body of Jesus, which is of the same substance as the Church."[151]

By cutting through the upper and lower realms with its vertical stave, cutting across the same division that its crossbar creates, the Cross makes possible the movement between realms, while also avoiding the collapse of one into the other. The vertical stave of the Cross/Limit/Redeemer acts not only to divide right from left but to collect, to gather toward and together by crossing another dividing line. The constant, never quite stable interactions of collection and division, one and many, whole and part appear over and over, generating and sustaining. Publication of the book-body on the cross is required for reading, but the reader is a member of the body and a name in the book, and is thus the book and the body too, and can know this once publication has happened, bringing body and book to the senses. We recognize here the double functions of Limit as separator and sustainer, the functions it performs from the establishment of the All onward.

The teacher of reading has to be brought before the senses and has to show through the senses both the power of Limit and the way across it. The Limit, as the place that is neither collected nor divided, as the source of creation and revelation, of cutting-off and unknowing, cannot belong to knowledge. And yet knowing can know *that* the Limit is. Because Limit is a power of the Son, the cross cannot be fully distinct from the body stretched on it, nor that body from the members of it, nor those members from the names written in the book that the body is. As all those who understand become members of the body, so too those written in The Living Book of the Living are the letters of a very peculiar name. Knowing the strangeness of the name and the text will mean knowing something both of the body of which those in the church are members and of the flesh that is the flesh of the world.

Holographic Text

I am the knowledge of my name.
The Thunder, Perfect Mind

Names are powerful in many religious traditions and are often associated with creation—a thing comes to be by virtue of being named. Genesis does not work exactly like this, but the God there does create both by dividing and by saying, and after creating gives names to what has been created: "God called the light Day, and the darkness he called Night."[152] Names grant a certain reality to the things that are made. Names may also be transformative. Theodotus describes the Valentinian baptismal rite as the receipt of an angelic name, pairing the human soul to its companion angel even as it joins the member of the church to the greater body. Like the Aeons, each angel is an aspect of what Brons calls "the dynamic richness of Jesus."[153] Names take the baptized across limits, allowing them, like redeemed Wisdom, to rejoin the All, the realm of the angels. Theodotus writes, "When we, too, have the Name, we may not be hindered and kept back by the Limit and the Cross from entering the [All]. Wherefore, at the laying on of hands they say at the end, 'for the angelic redemption' that is, for the one which the angels also have, in order that the person who has received the redemption may be baptised in the same Name in which his angel had been baptised before him."[154] Human and angel are joined by being given a joint name. They are given that joint name by a touch between bodies, the laying on of hands, bringing the word and the flesh together once more.

We read in the *Gospel of Truth* about a revelation given to the Aeons, the same revelation written in the Living Book. This is what those who learn to read can also come to know.

> He revealed it as a knowledge that is in harmony with the expressions of his will—that is, knowledge of the living book, which he revealed to the eternal realms at the end as his [letters]. He showed that they are not merely vowels or consonants, so that one may read them and think them devoid of meaning. Rather, they are letters of truth; they speak and know themselves. Each letter is a perfect letter like a perfect book, for they are letters written in unity, written by the father for the eternal realms, so that by means of his letters they might come to know the Father.[155]

In this book, all readers find themselves. But even the writing is extraordinary. The book is perfect in each truth-conveying letter, a book in every part, which does not, though, cease to be a part. The letters, into which a word can be divided, are written in unity. Each member of the body reads each letter that is the whole book; like a book of such letters, the body is complete in each member, though no member knows the whole of the sacred flesh. Once more one and many can neither converge nor be pulled apart. Names, as Patricia Cox Miller reminds us, aid our understanding, but "do not constitute or in any way circumscribe that Presence: 'our teaching is of the road and the travelling.'"[156] She writes of the *Tripartite Tractate* that it "focus[es] on language as a quest that both conceals and reveals meaning."[157] It is in this language that the names in the book are written.

Among the ways of naming the unsayable, one is especially powerful. In the *Gospel of Truth* we read, "The Name of the Father is the Son."[158] This name is paradoxical, at once unsayable and revealed—"The Father's name is not pronounced; it is revealed through a Son."[159] The *Gospel of Philip* plays on this multiplicity: "Jesus is a hidden name," it says; "Christ is a revealed name. . . . Nazarene is the revealed form of the hidden name."[160] Like the body accessible to the senses, the body that publishes the Book on the Cross, the name that is said in the world is a way of revealing the unrevealable, revealing that there is more than can be revealed. What is written in the book reveals the Father. To reveal him in writing requires that all of the names in the book make the name of the Father, a revelation of the unrevealable that does not alter its necessary hiddenness. Nor its incomprehensibility, as each letter says the whole of the book, the name of truth. It says the Son, the Name of the

Father, who is also a gatherer: of the Aeons, of the church members, of the letters.

According to Irenaeus, Valentinus's disciple Marcus amplifies upon this theme by identifying the Aeons themselves with the letters of the name. Letters do not know what they spell. Similarly, "there is not one of them that perceives the shape of that [utterance] of which it is an element."[161] Like us, the Aeons are ignorant of the full Name—if they knew it, they would know the unfathomable source, all succumbing to the danger that necessitated Wisdom's division. They are both single letters and complete truths, but they do not know themselves as completeness.[162] The truth is one that can only be what it is by being many letters. Like the Aeons as aspects of Christ and the church members as parts of the body, the letters together form the Son, the redeemer, the Name by which the Truth is known. The Name teaches reading, and what is read is the Name. The Name is made up of all the names written in the book. Each name in the book is made up of letters that themselves are each name and the whole Name. The Name of the Father is both complete and incomplete, collected into each letter and divided across all the letters. Matter, similarly, is shaped into the forms of the Aeons and the uncreated: in it, too, divinity is at once part and whole, reflecting an unimaginable ultimate source. Those who learn to read, if they can make out their own names in the Living Book of the Living, will learn to taste and touch and smell, to see and hear, the book that the world is, with its letters no less divine names.

The Unfinished: The Matter of Fractal Divinity

> Infinite-Limited, is it you?
>
> Maurice Blanchot, *The Writing of the Disaster*

In the *Gospel of Truth*, we read, "From the moment when the father is known, deficiency will cease to be."[163] This is not because everything has become indifferently collapsed or merged into one, nor because all that could be is filled, but because every fragment, without ceasing to be a fragment, is also complete—the name in each letter, the body in each member, the image of the All in each shape of matter. The mode of truth here, the way in which the Son as Name and Body is the image of Truth, can be read as holographic. That is, in each bit, the whole image is present: the whole body in each member, the whole book in each letter, the whole Name in each name. A smaller piece of a hologram gives us a somewhat smudged, less vivid image than a larger piece does, but each fragment "contains" the whole image nonetheless; each piece is also perfect, without deficiency. What makes this particularly fascinating here is that the image can only be what it is insofar as it has each part; it is not the image of an undifferentiated blob. The whole and part, the unified and the multiple and the undivided, are each within the other.

But a perhaps still more accurate metaphor is the fractal, as it is more mobile than the holograph. The "completion" implied by the lack of deficiency never means being finished and becoming still. Completeness does not rest, because the truth in every bit is a paradox, and paradoxes are mobile, and because divine love is sustained not by dissolution, but by infinite progress. The play of Limit and Unlimited continues to work generatively.

Edward Moore argues that gnostic thought is actually hostile to cosmic stasis: "When the disruption, brought about by the desire of Sophia, disturbed the Pleroma, this was not understood as a disturbance of an already established unity, but rather as the disturbance of an insupportable stasis that had [erroneously] come to be observed as divine."[164] Wise disruptiveness restores the movement required for generativity, as delimitation creates the Aeons so that they may continue to desire. The arithmetical world unfolding from the One and Unlimited Dyad is always in process, always creating. It generates in wholeness and division. When the All gather to form the name and image of the father, they participate not in finality but in creation. Creativity continues throughout creation. This creativity places, in Kenny's words, "divine fecundity and plenitude at the center of its understanding of divinity."[165] Given the arithmetic foundations of so much of this thought, we might consider this fecund movement as fractal.

Though they lack precise definition, fractals are roughly curves that, like other curves, can be made and described by mathematical formulae. Each result is plugged back into the formula to generate the next result. The formulae themselves can be pretty simple. The results, though, quickly become quite complex, even apparently chaotic. Each new step packs the same complexity or multiplicity into a smaller space; as Christoph Bandt explains, "When we see a piece of the figure, we cannot conclude where we are, nor can we say something on the size of the piece."[166] And each miniscule part, magnified to visibility, is the image of the whole. The better-known fractals, such as snowflakes or ferns, are clearly self-similar on every scale, but even the most visually difficult repeat their all-the-way-down roughness and curious dimensionality at every level. One can zoom in on a mathematical fractal forever. Their dimensionality, accordingly, is not quite the whole numbered version of our ordinary experience. And when a fractal curve encloses a space, such as that of the Koch snowflake, that space is a finite area delimited by an infinite boundary.[167] The Limit, unlimited by anything else, gives the form that makes perception possible—perception of the very infinite that we cannot perceive.

Some of the earliest work on fractals has to do with efforts to describe difficult natural forms such as jagged coastlines. Benoit Mandelbrot, the mathematician who coined the term "fractal," writes, "What a mad combination of contradictory features! Have we not finally come to mathematical monsters without conceivable utility to the natural philosopher? Again, the answer is emphatically to the negative." In fact, he says, what appears initially as highly abstract theory offers "the precise tool" needed to grasp, mathematically, "the geometry of . . . flesh."[168] (The ellipsis is his and suggests dramatic emphasis rather than missing words.) He notes the patterns of blood vessels and the branching of bronchia in the lungs, adding exuberantly, "Lebesgue-Osgood monsters [a particular kind of fractal curve] are the very substance of our flesh!"[169]

Mandelbrot's *The Fractal Geometry of Nature*, published in 1982, quickly acquired the status of a classic. In an essay published a dozen years later, Orly Shenker objects to the wide acceptance of Mandelbrot's claim that fractal forms are to be found in nature. While Shenker's essay includes empirical discussion, what may be most relevant for our purposes is the argument that "fractals are geometrically abnormal, in the sense that they are better viewed as geometrical processes, not geometrical objects," and are thus "ontologically and epistemologically . . . problematic."[170] For our purposes, though, this processual ontology and epistemic slippage are less problematic than promising. They suggest an unexpected neo-Pythagorean twist in contemporary mathematics.

In fact, even when we see a fixed and finished-seeming fractal shape, one that is drawn or found in nature, we understand its fractal character by considering the process that generates it. In some forms, such as those of fractal trees, that process is especially obvious; just by seeing the picture of the tree, we easily imagine the trunk dividing into two branches, each of those into two more, and so on.[171] At the beginning, there is one that is two; that is, one that folds and unfolds to delimit the unlimited in the image of both the one and the folded or divided. The simple-seeming image increasingly complicates itself, and the original curve becomes increasingly harder to perceive. But it never ceases to be that image. Within each part of it, the whole is spelled out, the body replicated, the form copied, though often at a scale that makes it illegible to those who do not know the formula. Kalvesmaki declares, "The Valentinian systems . . . are accounts of originate number, and of how reality springs from it. Valentinianism is, in part, a metaphorical expansion of Moderatus' second definition [of number] furnishing it with a vivid mythology."[172] Such mythology of lively, generative number offers us not an escape from the world, but the world's divinization—or, more exactly, a revelation of the world as already divine, in flesh and in name—that is, whether we experience its divinity as body or book. If the whole is in each, we do not deny the complexity and multiplicity of matter, but rather see the complex and the simple within each other. As the world unfolds, we read the impossible primal one and unlimited dyad at every level, in every image.

Obviously, a perfect, stable fusion of all of these opposites—sense and the imperceivable, completion and incompleteness, hidden and visible, one and two, nameless and named, whole body and member—would be impossible. We can now see that this is good. As Einar Thomassen reminds us, the Aeons engage in "an incessant teleological process of learning directed towards the Father."[173] He elaborates: "The [All] does not exist as a static structure but as a process, whose directing goal is knowledge of the Father, and unity, both with the object known and internal unity. But because the aeons are endowed with the freedom of will this unity must remain a potentiality and the process an unending one."[174]

The entire material world, with humans included, is generated out of the play of incompletion and completeness, collection and division, simplicity and complication—both epistemologically and ontologically. It is easier, complexity notwithstanding, to see the holographic and fractal quality of spirited humans who may learn and pass on learning, who read their own infinitely recursive names and form members of an infinite divine body. But just as the name and the body are images of an unimageable divinity, so too is the rest of the material world. Flowing forth from Wisdom and formed to reflect the collected Aeons, all of matter is thus an image of the recursively perfect letters that cannot read the name they spell, because those letters, says Marcus, are the Aeons themselves. All the world teaches, in a way that is most concentrated in the Savior's redemptive body—but that teaching is that all must be taught by all things. Beauty has meaning. Yet again the neatness of the active subject and the passively receiving object is made unstable. As Moore nicely summarizes, "The goal, on this view, is to produce through wisdom, and not simply to attain wisdom as an *object* or end in itself. Such an existence is not characterized by desire for an object, but rather by desire for the ability to persist in creative, constitutive engagement with/in one's own 'circumstance.'"[175]

Miller points out that the emphasis on desire as generative, which is an emphasis on Eros, was another source of hostility for the heresiologists:

> Expressed primarily in metaphors of desiring, lovemaking, and giving birth, Gnostic theological language has sensuous qualities that are striking. This did not escape Irenaeus, who at one point chooses to ridicule Valentinus's sexual vision of making with an equally organic

and sensuous language, not from the human but from the vegetative world, envisioning fruit "visible, eatable, and delicious." Valentinus's "melons" might be "delirious," but Irenaeus's choice of metaphor is revealing: he seems to have realized that Gnostic thinking about making had placed "the intercourse of Eros" at center stage.[176]

Hippolytus tells us that despite the peaceful quiet, "(the Father) was not fond of solitariness. For . . . He was all love, but love is not love except there may be some object of affection."[177] His first thought is Desire. The world comes to be in love thinking itself in order to have something to love. Desire and knowledge and the desire to know move only within difference. This is why the One, which imposes limit, creates difference in the process; this is why the unlimited Dyad has the power to receive limitation. The two work together in the completeness and incompletion of the world. When we read the world rightly, it speaks (or perhaps even sings) to us: we realize that reminders of divinity are everywhere, and we too can fall in love, can be enchanted and seduced by beauty, so that we can come to know. Redemptive knowledge does not take one out of matter, but grounds the human in it as part of it, part of the same set of holographic relations and fractal movements that make the world divine. The same thing is there for thinking of and for being. Thinking, by itself, is missing something; being, by itself, is incomplete. The limit of knowing encountering the limit of being reveals the inherent mystery, the traces of divinity, the creativity of matter.

Gnosis is thus a complicated revelation. Epistemologically, it is a matter of recognizing limit as at once a barrier and a revelatory site. The ontology that it recognizes in that revelation is that of a lively, mobile, and divine materiality; a matter that is not other than the knowers but in which they are caught up—matter not as something to be overcome, but as a way at every level to encounter the wisdom that tells us that there is more than knowledge. Redemption must be presented to the senses, presented by a meaningful body to teach that no flesh is wholly delimited, that divinity is neither lacking nor finished. Because knowledge and desire need each other, the world cannot be simply divided into agents and objects; what is desired acts upon the desiring one every bit as much as vice versa. Such knowledge pulls on memory. Matter madly multiplies both as a mnemonic and as the very beauty at its source. No disembodied redeemer could teach such complication. Matter itself harbors mystery: every moment replicates its firstness, and to replicate divinity means that the world was always already redeemed, always divine even in its sorrow. We only forgot how to see it. We are reminded to be astonished.

4
Glorious Return
Resurrected Bodies

Affirmations of Return

> What the Mystery reveals, therefore, is *the body as revealed mystery*, the absolute sign of self and the essence of sense, God withdrawn into flesh, flesh subjectivized to itself, which finally, is called "the resurrection," in the full radiance of the Mystery.
>
> Jean-Luc Nancy, Corpus

In the end, the body begins anew. Like the primal and salvific bodies, the glorious body of the resurrection opens onto the flesh of the world to reveal, as Paul writes to the Corinthians, a mystery: "Listen, I will tell you a mystery! We will not all die, but we will all be changed, in a moment, in the twinkling of an eye, at the last trumpet. For the trumpet will sound, and the dead will be raised imperishable, and we will be changed."[1] Resurrection starts over and better. It offers an improved version of what is currently imperfect.

There are even versions of a kind of cosmic resurrection, world after world. We read in *Genesis Rabbah*,

> There was a time-system prior to this. Rabbi Abbahu said: This teaches us that God created worlds and destroyed them, saying, "This one pleases me; those did not please me." . . . Rabbi Abbahu derives this from the verse, "And God saw all that He had made, and behold it was very good," as if to say, "This one pleases me, those others did not please me."[2]

Though Isaac Luria differs from some other Kabbalists by not proposing a series of created worlds, a later Kabbalistic variant connects this midrash to his cosmogony, in which the divine light is so powerful that it shatters the vessels that were meant to contain it: "The worlds that were created and those that were destroyed were the shattered vessels that God had sent forth. Out of those broken vessels God created the present universe."[3] Earlier than either of these versions of re-creation, the Stoics held that the cosmos is infinitely remade just the same as it ever was; the same, because it embodies divinity in its very principles, and so its perfection cannot be improved upon. (They attributed this cosmogony to Heraclitus, but this was probably an over-interpretation on their part.) And before them, the Pythagoreans posited a grand recurrent cosmic cycle of identical events.

Though it is more obvious in these all-encompassing cosmic recurrences, all notions of embodied resurrection disturb the meanings of "beginning" and "end." Ends become new beginnings. Of course, neither the person nor the cosmos glides in a smoothly returning circle. The boundaries of the dusty and luminous microcosmic Adam and the sensuous limiting savior Christ turned out not to hold steady after all, but in their different ways to engage all of matter with all the rest and to scatter mysterious divinity throughout. So too does the matter of risen bodies entangle with the material of the world. And like the microcosmic boundaries of the first human, like the divisions, enclosure, and disclosings of divine limit, the edges of this near-circle of ends and beginnings can neither meet up nor hold steady. The where, the when, and the what of risen bodily stuff all evade neat and reasonable positioning.

This difficulty has not deterred many different kinds of thinkers from asserting that flesh rises and that believing so is important. A Talmudic passage assures the reader that punishment awaits those who disbelieve the reality of resurrection: "He denied the resurrection of the dead; therefore he will not have a share in the resurrection of the dead, as all measures dispensed by the Holy One, . . . are dispensed measure for measure, i.e., the response is commensurate with the action."[4] On the one hand, this response does not seem commensurate at all—infinite inexistence for a mere opinion. On the other, it is nothing but confirmation—one who disbelieves in resurrection will be proven right. In any case, the suggestion that some might deny any sort of resurrection was not a product of rabbinic imagination. In the first century of the common era, historian Flavius Josephus reports that the Sadducees were especially skeptical, while the Pharisees (Josephus himself included) affirmed resurrection.[5] The Talmud describes the Saducean skepticism: "The Sadducees questioned Rabban Gamaliel: 'Whence do you infer that the Holy One, praised be He! would restore the dead to life?' And he answered: 'From the Pentateuch, Prophets, and Hagiographa.' However, they did not accept it."[6] In Christian scriptures, all of the synoptic gospels and the Acts of the Apostles make similar though less detailed claims for Jesus, presenting him as likewise being questioned by Sadducees, who "say that there is no resurrection."[7] And Paul, calling himself "a Pharisee, a son of Pharisees," puts himself on the Pharisaic side in the dispute: "The Sadducees say that there is no resurrection, or angel, or spirit; but the Pharisees acknowledge all three."[8]

According to Josephus, a third Jewish group, the ascetic Essenes, believe in a resurrection, but not of the body (though there is now some doubt about his accuracy here).[9] Evidence for the Pharisees' beliefs regarding bodily resurrection is stronger, but still ambiguous. A clear affirmation of bodies that rise will wait for the Christians and for the rabbis, who are largely the Pharisees' intellectual descendants. As Daniel Silver points out, though, "Resurrection [is] affirmed rather than defined by the rabbinic tradition,"[10] and certainly Paul did not make the matter any clearer. *What* resurrection might be remains a subject of contention, even with the affirmation *that* resurrection is.

Both rabbinic and early Christian exegetes read that affirmation back into scripture as having always been fundamentally important. In a third-century midrash, we read, "There is no section of the Torah that does not deal with the resurrection, but we lack the strength to expound it."[11] A Talmudic text from about the turn of the sixth century emphasizes the necessity of resurrection for good action: "There is no mitzvah written in the Torah that you are given its reward without depending on the resurrection of the dead."[12] To believe in the resurrection of the dead is both necessary and rewarding, or at least implicit in every reward.

The scriptural passages used to support claims of resurrection are varied. The sixteenth Psalm declares life and joy sempiternal: "In your presence there is fullness of joy; in your right hand are pleasures forevermore."[13] Other passages emphasize the divine ability to heal, even from death: "The Lord kills and brings to life; he brings down to Sheol and raises up"; "I kill and I make alive; I wound and I heal"; or, from Qumran, "For He shall heal the critically wounded, He shall revive the dead."[14] Though arguably an abstract soul or spirit might be wounded and healed, other scriptural indications are more emphatically corporeal. We read in Isaiah, "Your dead shall live, their corpses shall rise. O dwellers in the dust, awake and sing for joy! . . . The earth will give birth to those long dead."[15] Job too speaks of skin and flesh as what is to be destroyed and then redeemed.[16]

One of the passages most cited as evidence for bodily resurrection comes from Ezekiel. There the prophet's speech works together with divine intervention to reassemble bare scattered bones and give them both flesh and breath. The process of reassembly is presented in some detail:

Suddenly there was a noise, a rattling, and the bones came together, bone to its bone. I looked, and there were sinews on them, and flesh had come upon them, and skin had covered them; but there was no breath in them. Then he said to me, "Prophesy to the breath, prophesy, mortal, and say to the breath: Thus says the Lord God: Come from the four winds, O breath, and breathe upon these slain, that they may live." I prophesied as he commanded me, and the breath came into them, and they lived, and stood on their feet, a vast multitude.[17]

Canonical, orthodox Christianity, following the Pharisee Paul, is emphatically in favor of bodily resurrection—first claiming it for Christ, and then (consequently) for the rest of humanity. The claim that some Christian sects disavow the reality of bodily resurrection, mirroring the claim that they deny the reality of Christ's body, is grounds for labeling them heresies—though as we must already suspect, that claim is not always unambiguously supported. Paul is emphatic: "Now if Christ is proclaimed as raised from the dead, how can some of you say there is no resurrection of the dead? If there is no resurrection of the dead, then Christ has not been raised; and if Christ has not been raised, then our proclamation has been in vain and your faith has been in vain."[18] Paul hopes for and evidently presumes the return of Christ within his own lifetime. Christ's second coming—or rather, a second return after his resurrection into and departure from this world—portends the bodily resurrection of humanity.

Other New Testament passages declaring the resurrection of Christ are too numerous to list, appearing in all of the canonical gospels, the Acts of the Apostles, the book of Revelation, and many of Paul's letters. Though less numerous, references to a more general resurrection are still plentiful. Paul's, again, are the best known and probably the most influential. Though insisting on bodily resurrection, he hardly regards it as straightforward, as the already cited passage from Corinthians shows. "This perishable body must put on imperishability," he writes, "and this mortal body must put on immortality."[19] As Christian orthodoxy develops, and as it becomes clear that the end and renewal heralded by Christ's second coming are not arriving so quickly as Paul had thought, some details are affirmed in all their ambiguity: what will rise is indeed body (somehow), and it will rise at some future time (somewhen), which might then have the effect of ending the passage of time altogether (whatever "then" might mean without time). Irenaeus declares that Christians must believe "the resurrection from the dead, and the ascension into heaven in the flesh of the beloved Christ Jesus, our Lord, . . . to raise up anew all flesh of the whole human race."[20] Tertullian likewise makes resurrection central: "The resurrection of the dead is the Christian's trust. By it we are believers."[21] Tertullian deals with some of Paul's trickier phrases, notably, "Flesh and blood cannot inherit the kingdom of God," by declaring that it is not flesh in the sense of body, but rather the (bad) actions that we do in the flesh, that cannot inherit heaven.[22] Despite the efforts of the heresiologists to deny it, affirmations of resurrection also emerge among the "heretics": "But since you ask us pleasantly what is proper concerning the resurrection," writes the Valentinian author of the *Epistle to Rheginus*, "I am writing you that it is necessary."[23] The nature and sources of that necessity, though, remain obscure. The body that "will be changed" remains or rebecomes flesh, but it is transformed in some way that is not quite susceptible to saying, at least not straightforwardly.

Yes, But When?

> The disciples said to Jesus, "Tell us how our end will be."
>
> Jesus said, "Have you discovered, then, the beginning, that you look for the end? For where the beginning is, there will the end be. Blessed is he who will take his place in the beginning; he will know the end and will not experience death."
>
> *Gospel of Thomas*, logion 18

The proto-orthodox insistence on resurrection in the future is nearly as vigorous as the insistence on resurrection in the body. As we must already suspect, Christian salvation and resurrection are intimately connected concepts, and in this they understand themselves to be building on earlier Jewish ideas. Resurrection is the reward for salvation, and the nature of that reward necessarily differs depending on the understanding of what salvation is. The reward for redemption achieved through sacrifice, the usual understanding of Jesus as messiah, is set in the future. Irenaeus and Tertullian ground their arguments for futurity in Paul. Even before Christianity had any established orthodoxy, some Christians saw heretical possibilities in claims for a resurrection that was either past or present. In the Pauline letter to Timothy, the author condemns those Christians like "Hymenaeus and Philetus," "who have swerved from the truth by claiming that the resurrection has already taken place." It is not simply that they are mistaken themselves; "they are upsetting the faith of some."[24] We do not know much about those who are described as upsetting the faith, and as Dale Martin points out in *The Corinthian Body*, it is probably anachronistic to attribute to them a fully developed belief in a realized eschatology or an immanent salvation.[25] Still, this anxious criticism of those who "upset the faith" is similar to the remarks Irenaeus and Tertullian both will make regarding the "heretics," who did more clearly develop such ideas. Among other errors, these heretics suggest that resurrection might not be an event set to happen later.[26] For the heresiologists, resurrected bodies *will be*—not *are*—fleshy bodies. The wait is meant to favor flesh: in an oddly definite dualism, Irenaeus, Tertullian, and the author of the letter to Timothy all assume that an immanent resurrection can only be of the soul alone, and a bodily resurrection can only occur in the future.[27]

Sorting through scriptural passages regarding a final judgment, Augustine agrees that "the resurrection of the body . . . shall be in the end."[28] Yet in both *City of God* and his tractates on the Gospel of John, he attempts to interpret the strange doubleness of a sentence from that gospel's fifth chapter: "The hour is coming, and is now here, when the dead will hear the voice of the Son of God, and those who hear will live."[29] He tries to make sense of the double hour by positing a double resurrection. First is a present and somewhat metaphorical resurrection of the soul, which can rise away from at least some measure of its attachment to sin, though in this life we can never be wholly free from that attachment. The second, more enduring, and more dramatic resurrection is the body's, whereupon it is rejoined by the soul. The idea of double resurrection becomes a standard Christian doctrine. Thomas Aquinas will offer another version of it, arguing that souls live on and are immediately rewarded or punished upon death, and only later rejoined to their bodies. Between Augustine and Thomas, two important Jewish thinkers also posit multi-step resurrections, but of quite different sorts. Moses Maimonides, whom Thomas greatly respects and from whom he sometimes borrows, presents a first resurrection of body and soul together.[30] This is so that they might be judged, since, as the Talmud also argues, neither has acted without the other.[31] Thomas follows Maimonides in arguing that body and soul must be judged together, though for him and some earlier Christians this step comes later in the process.[32] For Maimonides, however, the final stage of resurrection belongs to souls alone, and only the souls of the righteous go on to the happiness of the world to come.[33] "The World to Come," he insists, "harbors neither body nor aught of concrete form, save only the souls of the righteous divested of body as are the ministering angels."[34] His first step, the joint judgment of body and soul, strongly resembles the one proposed by Saadia Gaon, who in the tenth century gives us the first full treatise on resurrection in Jewish thought. But Saadia's final stage differs; he proposes a perfection of ensouled bodies in a world illuminated by Wisdom. The bodies are resurrected in recognizable, lived form in order that they may be judged. They may then be perfected according to their goodness.[35] This perfection is as bodily as the judgment was, and it is in bodily perfection that the good inhabit the world to come. That world is a transfigured version of our world, as risen bodies are transfigurations of ours.

For Thomas, the souls that have been split from their bodies await the rising of those bodies at "the end of the world." The wait may be more or less pleasant, depending upon the souls' goodness.[36] He and Augustine agree, though, that perfect happiness can only come when body and soul reunite.[37] Though they seem to regard their own solutions as satisfactory, one cannot help suspecting that the author of John, who was fairly comfortable with mystery and mysteriousness, might be doing something more interesting than dividing resurrection in two. In fact, the sentence just before the one that Augustine analyzes could certainly be read as implying a more immanent and not merely imminent event: "Very truly, I tell you, anyone who hears my word and believes him who sent me *has* eternal life, and does not come under judgment, but *has passed* from death to life."[38]

A rather elegant variation on future embodiment appears in Augustine's *City of God*, and it complicates futurity itself. Despite being Platonic in so many respects, Augustine argues against the common Platonic idea that embodiment is the consequence of the soul's forgetfulness, its inattention to the gods. If anything, he suggests, the dead *remember* corporeality and rightly long for it:

> It is not, as Plato imagined, through forgetfulness that [souls] long to have their bodies again. In fact it is just because they remember the promise of him who never lets anyone down, who gave them the assurance that even the hairs would remain intact; remembering this, they look for the resurrection of their bodies with patient longing, for though they suffered much hardship in those bodies, they will experience nothing of the kind hereafter.[39]

Soul and body are properly joined in a state of joy. They remember the promise of this joy—in fact, Augustine argues that we know what happiness is because we find it in our memories, even though we never perfectly experience it in our lives.[40] The perfect human state is not one in which we are "divested of" corporeality.[41] Not only is the body better off if ensouled, the soul is better off if embodied: "And although we must never for a moment doubt that the souls of the righteous and devout live in a state of rest after their departure from this life, yet they would be in a better state if they were living in conjunction with their bodies in perfect health."[42] Despite Augustine's vehement orthodoxy, it is not hard to see how easily Valentinians might find themselves in agreement with redemption-by-remembrance, not just as an intellectual matter but in the flesh.

In keeping with this loop that seems to draw what is gone and to come through the present, the traditions that include bodily resurrection also link it to creation. Re-arising is possible from the bodies' first making. In some fashion, resurrection is what flesh is made for.

Created for Resurrection

> Human beings perish because they are not able to join their beginning to their end.
>
> Alcmaeon of Croton, Fragment 2

The very idea of a divine creator is sometimes used to argue for God's miraculous power over death. To create at all, these arguments say, shows power over chaos or even over nothingness. Resurrecting bodies is a comparatively minor miracle. Athenagoras finds it absurd to say of a God who has made flesh that he cannot remake it; "the 'restoration' of the flesh is easier than its first formation."[43] Irenaeus argues the same:

> God, taking dust from the earth, formed man. And surely it is much more difficult and incredible, from non-existent bones, and nerves, and veins, and the rest of man's organization, to bring it about that all this should be, and to make man an animated and rational creature, than to re-integrate again that which had been created and then afterwards decomposed into earth.[44]

Tertullian asserts that whether God has created *ex nihilo* or from preexistent matter, he has brought things into existence that did not exist before, and it would be absurd to think that he could not then remake them from their materials.[45] Augustine likewise points out how easily a God who creates everything out of nothing could go on to perfect bodies in the resurrection.[46]

Similarly, we read in the Talmud the story of the daughter of a skeptical emperor, who explains the possibility of a bodily resurrection through an analogy: A craftsman who can make a vessel out of mortar is good, she says, but one who can make a vessel out of water is extraordinary. The God who creates the world from the formless depths, then, could easily perform the lesser creative act of molding clay as well. And so, "If God was able to create the world from water, He is certainly able to resurrect people from dust."[47] Another passage there offers a different metaphor for the relative ease of recreating a body:

> If concerning glass vessels, which are fashioned by the breath of those of flesh and blood who blow and form the vessels, and yet if they break they can be repaired, as they can be melted and subsequently blown again then with regard to those of flesh and blood, whose souls are a

product of the breath of the Holy One, Blessed be He, all the more so can God restore them to life.[48]

Saadia repeatedly reiterates this point, arguing that resurrection is surely easier, as act and as concept, than creation. The divine capacity displayed in the initial creation is evidence in itself for the possibility of re-creation.[49]

Several midrashim on the Ezekiel story offer an anatomically specific version of a body made for remaking. They go beyond the broad reference to "bones" to refer more narrowly to the "resurrection bone" or "luz bone," a small bone at the end of the spine "from which God will one day cause man to sprout forth again."[50] Potentially, then, a body harbors the principle of its own recurrence, not as an event imposed from without, but as that body's own fulfillment. This idea is prominent in Kabbalah.[51] The *Zohar* uses the image of yeasted bread: "For one bone remains from the body beneath the earth, and that one never decays or wastes away in the dust. At that time, the blessed Holy One will soften it and make it like leavened dough—rising, expanding in four directions, thereby forming the body and all its smooth members. Afterward, the blessed Holy One will infuse spirit into them."[52]

Other authorities, including the midrash *Sifrei Devarim* and the apologetics of both Irenaeus and Tertullian, link the revivification in Ezekiel to the second story of human creation in Genesis, where "the Lord God formed man from the dust of the ground, and breathed into his nostrils the breath of life; and the man became a living being."[53] In both stories, the dust is gathered together into flesh and enlivened by making that flesh breathe a breath it shares with God—and, in Ezekiel, with the winds of the world. As a second creation, resurrection repeats the processes of the first. In Isaiah, the dust is made lively, and not dry earth but living nature arises: "Your heart shall rejoice, and your bones shall flourish like the tender grass."[54] The body's rising emphasizes its earthiness, rather than disconnecting it from the rest of material creation—the earthy body flourishes again, beginning with its bones.

Tertullian, surprisingly for a Christian polemicist, evokes the seasonal return so foundational to ancient mysteries. In so doing he thoroughly entangles human resurrection with the matter of the world. "Day dies into night and is on every side buried in darkness. . . . And yet again the same light . . . revives for the whole world, slaying its own death, the night, stripping off its funeral-trappings." Likewise, there is "a returning home" of constellations and of the phases of the moon; "a revolution of winters and summers, of springs and autumns, with their own functions, fashions, and fruits." This leads Tertullian to the remarkable declaration that "all creation is instinct with renewal. Whatever you may chance upon, has already existed; whatever you have lost, returns again without fail. All things return to their former state, after having gone out of sight; all things begin after they have ended; they come to an end for the very purpose of coming into existence again. Nothing perishes but with a view to salvation." This "whole . . . revolving order of things," he declares, "bears witness to the resurrection of the dead."[55] All things that are made are made for remaking.

The *Zohar* also connects the resurrection of humans to that of nature, but in a quite different way. Telling the story of the transformation of Aaron's staff into a blossoming branch, it argues for the ease with which God can gather the dust into newly living beings: "Later, when he joyously renews the world, how much more so will he transform them into new creatures!"[56] The story itself, from the book of Numbers, moves quickly from the budding of a stick of dead wood to being spared from death, though that sparing might not be an everlasting one:

> Moses . . . saw that Aaron's staff . . . had not only sprouted but had budded, blossomed and produced almonds. Then Moses brought out all the staffs from the Lord's presence to all the Israelites. They looked at them, and each of the leaders took his own staff. The Lord said to Moses, "Put back Aaron's staff in front of the ark of the covenant law, to be kept as a sign to the rebellious. This will put an end to their grumbling against me, so that they will not die.[57]

Not all readings of Genesis, we may recall, attribute creation to a solitary god acting unassisted. Nachmanides's suggestion of earth as a co-creator fits nicely with the idea of a risen body remade of earth again—though of a transfigured earth, as we shall later see.[58] Besides the possibility of angelic co-creators, *Genesis Rabbah* also offers that of the creator working with the earth and heavens: "'Let us make a human'—with whom did He rule? R' Yehoshua in the name of R' Levi said: With the work of the heavens and the earth."[59] This pairing is the necessary condition for the possibility of resurrection. The creator declares, "If I create him of the celestial elements he will live [forever] and not die, and if I create him of the terrestrial elements, he will die and not live [in a future life]. Therefore I will create him of the upper and of the lower elements: if he sins he will die; while if he does not sin, he will live."[60] According to Rabbi Simai's similar description in the *Sifrei Devarim*, God calls "to the heavens from above and to the earth" to "deliberate with him in judgment" as to the resurrection of the person, since the person's soul is from the heavens and body from the earth, and these always act together.[61] We may also recall Jacob Asher's variation on this idea, in which the earth, once created, further participates in making bodies.

In the pseudepigraphic text of 2 Baruch (from the first or second century CE), though the earth is not assigned an initial creative role, the Lord says that it acts with him in raising the dead. "The earth shall then assuredly restore the dead, [Which it now receives, in order to preserve them]. It shall make no change in their form, But as it has received, so shall it restore them, And as I delivered them unto it, so also shall it raise them."[62] The Lord will go on to transform that flesh, but the earth is active and cooperative in first restoring and raising it. *Genesis Rabbah* even offers an optimistic reading of the rather grim line in Genesis, part of the humans' punishment for disobeying their God—"You are dust, and to dust you shall return."[63] In this interpretation, "scripture hints at resurrection, for it does not say, For dust thou art, and unto dust shalt thou go, but *shalt thou return*."[64] The scattering of flesh into dust also entails a regathering of that dust, a return to its human shape.

This verse in Genesis is part of the story of the humans' exile from their perfect garden world, and in general it seems clear that something must happen to creation for a second beginning to be necessary. Whether the fault is a matter of forgetting or of disobeying varies across traditions. But the lapse is never final—the possibility of restoration is built into the beginning, available before any breaking has occurred. We might recall the way that God drew dust from all over the world, so that the disintegrated flesh of the mortal human might be at home wherever it is carried. Or the claim in *Genesis Rabbah* that return and reunion are possible even after wrong action: "The Holy One, blessed by He, opened an opening of repentance for him."[65] Christianity follows the traditions of Wisdom by having the one through whom the world is made reappear as the one by whom it is saved, whether by sacrifice and example or by shaping and teaching. The materiality of creation helps to insure, and to make sense of, a material resurrection.

Even where the role of a savior is pedagogical, the teacher reaches back to or before the beginning, bringing, as the *Tripartite Tractate* has it, "instruction and a return to that which they had been from the beginning—that of which they possessed a drop inciting them to return to it—which is what is called redemption."[66] In the *Gospel of Thomas*, Jesus tells his followers, "What you look for has come, but you do not know it."[67] The anticipated future is present, and in that present, those who understand return to "the kingdom. For you have come from it, and you will return there again."[68] The author of the *Gospel of Philip* clearly has a resurrection in mind, though just as clearly a strange one, in claiming, "Those who say that the master first died and then arose are wrong, for he first arose and then died." Far from simplifying the bizarre time of the resurrection as the culmination of a series beginning with birth, then continuing into life, then followed by death, this "rising up" from death precedes death itself, in or before the beginning.

Perhaps the most literal return of the end to the beginning is *apokatastasis*, the redemption of the whole cosmos that is its return to its state at the moment of its creation. It is enough of a variation to merit a more extensive discussion, and we will come back to it in looking at the particular resurrection of bodies that it implies. In considering its link to creation, however, we may note that the term makes a rare and somewhat vague scriptural appearance in the Acts of the Apostles, where Peter urges his listeners to repent, "so that times of refreshing may come from the presence of the Lord, and that he may send the Messiah appointed for you, that is, Jesus, who must remain in heaven until the time of universal restoration (apokatastasis)."[69] In the fourth century, Gregory of Nyssa argues for bodily resurrection leading to a final perfection in which "every will rests in God," fully embodied, and "God will thus be 'all in all,'" with every being restored to its original state, the one that preceded the division that disobedience caused.[70] A bit later in the fourth century, Jerome, commenting on Paul's letter to the Ephesians, brings this restoration together with the microcosmic Adam and the multimembered Christ: "In the end and consummation of the Universe all are to be restored into their original harmonious state, and we all shall be made one body and be united once more into a perfect man and the prayer of our Savior shall be fulfilled that all may be one."[71] Apokatastasis appears as well in the *Gospel of Philip*, where we read, "There is rebirth and an image of rebirth, and it is by means of this image that one must be reborn. What image is this? It is resurrection. Image must rise through image. By means of this image the bridal chamber and the image must approach the truth. This is restoration (apokatastasis)."[72]

Though it shows up in the work of all of these thinkers, universal restoration is most strongly associated with Origen, who held that bodies are gradually perfected such that they are always appropriate to their gradually improving souls. Before spatial and temporal things are created, Origen says, their (perfect) forms are eternally contained within Wisdom, to which they will be returned.[73] We may recall that for Origen bodies are made to "adorn the world"; material things are necessary both for the world's beauty in accordance with Wisdom and so that the souls might have a place to exist. Accordingly, restoration does not rid us of bodies; it occurs only when bodies and souls are perfected together. Salvation occurs through the same source as creation, which is Wisdom or Christ.[74] Like Gregory, Origen understands this restored state to be the referent of Paul's claim "that God may be all in all."[75] All things participate in God, not just people; all stuff is redeemed and is necessary for any possible redemption.[76]

In Kabbalah, restoration is a project in common for all. Here too resurrection works toward an ultimate perfection that restores a perfect primal unity. As is also the case for Origen, souls will be successively embodied, in ways that we shall explore shortly, and are ultimately reunited with their embodiments as part of the regathering of divinity. In this regathering, worlds are mended. This mending is *tikkun*, an imperfectly translatable term indicating repair, correction, or justice—an emphasis that distinguishes it from apokatastasis, which lacks the dimension of social ethics. In rabbinic texts we find the term used in the last of these senses to describe the rules governing social justice, particularly for the less advantaged.[77] Like so many ideas, this one takes on considerably greater complexity in Kabbalah. Most simply, it depends upon the mirroring of divine and mundane realms, so that good and just action in the everyday world also helps to mend the damage in divinity, drawing each together and the two into one another.[78] No act is isolated: "There is no good or evil, holiness or defilement, without its root and source above," says the *Zohar*. "By an action below is aroused an action above."[79] Eventually all divinity is restored, and there is no division between upper and lower worlds, between mundane and divine. Each good act helps to mend the world; each act of mending heals a rift with the divine realm. And each soul rises as part of this cosmic perfection together with its perfected body.

Clearly, there are many different versions of a resurrection built into creation, and they take in varying parts and proportions of the created world. In their different ways, they raise the persistent question of Paul's "mystery": if the created body is made for resurrection, what kind of strange flesh (re)arises? If resurrection returns an end to a beginning, must it be set in a future? Can it be set in a future? How strange must flesh be in the first place, for that resurrection to be possible? Where are its limits? Among the answers offered, some are particularly intriguing: that the flesh is already a part of a world full of divinity; that resurrection itself recurs because the soul occupies a series of bodies; and conversely, that the singular risen body gathers into itself all the matter that it has been over its life.[80] Each of these merits a more detailed exploration.

Curious Forms 1: If You Know How to Listen

> Rabbi Yehoshua ben Levi said to Elijah: The Messiah lied to me, as he said to me: I am coming today, and he did not come. Elijah said to him that this is what he said to you: He said that he will come "today, if you will listen to his voice" (Psalms 95:7).
>
> Sanhedrin 98a

Among the least straightforward texts on resurrection are two Valentinian sources, both of which we have already noted: the *Epistle to Rheginus*, sometimes called the *Treatise on Resurrection*, and the *Gospel of Philip*. Thoughtful scholars read these confusing works differently, with some finding and others refusing a resurrection of the body in them.[81] The *Gospel of Philip* rejects those who claim that an unchanged flesh will rise. But it also rejects those who hold that the soul will rise without flesh: "Both views are wrong," it declares.[82] The author demands,

> You say that the flesh will not arise? Then tell me what will arise, so we may salute you. You say it is the spirit in the flesh, and also the light in the flesh? But what is in the flesh is the word, and what you are talking about is nothing other than flesh. It is necessary to arise in this sort of flesh, since everything exists in it.[83]

Yet that same author also calls the soul "something precious . . . in a worthless body," and adds, citing Paul, "Some people are afraid that they may arise from the dead naked, and so they want to arise in flesh. They do not know that it is those who wear the [flesh] who are naked. . . . 'Flesh [and blood will] not inherit God's kingdom.'"[84]

Jorunn Buckley writes of the text's dizzying discussion, "The dualism presented in the *Gospel of Philip* is, essentially, only an apparent one; it can, the text instructs, be overcome by means of right insight and correct action."[85] Not matter, but duality, is an illusion. The confusion that the gospel generates depends upon thinking of body and soul as distinct. The *Epistle* too considers things that are illusory, noting how absurd it would be to deny resurrection: "Do not think the resurrection is an illusion. It is no illusion, but it is truth! Indeed, it is more fitting to say the world is an illusion rather than the resurrection."[86] Possibly this suggests that the world is an illusion and the resurrection is not. But possibly, more interestingly, the claim that the world is an illusion is not fitting at all: it is absurd, and it is even more absurd to deny resurrection.

As Buckley also notes, the overcoming of illusion must occur as we live. She points out that the *Gospel of Philip* "states that one must acquire divinity while still on earth; 'If one does not first attain the resurrection will he not die? As God lives, he would be (already) dead.'"[87] The resurrection is not an illusion; it happens in the flesh, *and* it must happen in one's lifetime. Resurrection still follows upon redemption, but the two are less distinguishable here than in proto-orthodox and then orthodox Christianities: resurrection is part of what salvation means. Resurrection is immanent to salvation rather than consequent upon it, a claim that will upset the faith of some.

The ambiguity regarding the flesh persists. Having asked pleasantly about the resurrection, Rheginus is urged not to "think in part . . . nor live in conformity with this flesh for the sake of unanimity," an instruction that resonates with less positive calls to abandon worldly joys. Then his correspondent continues, "but flee from the divisions and the fetters, and already you have the resurrection." Neither division nor unity matters; or rather, neither is to be simply affirmed.[88] Here we can join this Valentinian discussion of resurrection to the Valentinian sense of matter shaped by creative and salvific Wisdom and Limit. If what we learn from Limit is the infinite gathered-and-scatteredness of holographic and fractal divinity in matter, then to regard "oneself" as risen, to remember in the flesh, must open too upon the flesh of the world, telling us that it is already risen, or redeemed, or divine.

In distinction from proto-orthodox Christian descriptions, and in keeping with the later Kabbalistic descriptions, gnosis-oriented Christianities give the one who is resurrected considerable agency in the act of resurrection. In the proto-orthodox cases, the agency of an ordinary person in the face of resurrection lies in the past. At the time of resurrection, it is only relevant as the object of judgment; the same seems to be true in rabbinic texts, whether or not they endorse an ultimate resurrection of the body. The *Gospel of Philip*, on the other hand, takes its readers through a complex series of sacraments in which the distinct agents who redeem and who are redeemed become increasingly blurred. In the ultimate sacrament of anointing, one comes to possess all that the anointed one does: "The Father anointed the Son, the Son anointed the apostles, and the apostles anointed us. Whoever is anointed has everything: resurrection, light, cross, holy Spirit."[89] When the transfigurative power of the sacraments is truly known, "This person is no longer a Christian but is Christ."[90] "The master" therefore declares, "I have come to make [the lower] like the [upper and the] outer like the [inner, and to unite] them in that place."[91]

When a person receives the truth, *Philip* concludes, "the world *has become* the eternal realm."[92] Christ here literally shows his followers a mystery, a sacrament (*mysterion*).[93] In it, the entire world becomes (and has already become) sacramental—through this sacramental revelation, we see the truth in the image that the world is. The *Epistle to Rheginus* urges the same transformation through recognition: "If he who will die knows about himself that he will die . . . why not consider yourself as risen and (already) brought to this?"[94] Just as we ordinarily know that we will die, so too we can know that we are resurrected. The sacraments work with gnosis, and their working is theosis. "Nothing," the letter writer says, "redeems us from this world." The problem is not the place, but the preposition; we are redeemed not *from* but *in* this world. We recognize that "we are saved. We have received salvation from end to end. Let us think in this way! Let us comprehend in this way!"[95]

And this way is not unfleshed. When "the [master rose] from the dead," the *Gospel of Philip* explains, "[he did not come into being as he] was. Rather, his [body] was [completely] perfect. [It was] of flesh, and this [flesh] was true flesh. [Our flesh] is not true flesh but only an image of the true."[96] Our unrisen flesh is only an image of this truth, presumably awaiting truth itself for its transformation. The *Epistle* reminds us, in passages reminiscent of the *Gospel of Truth*, that we know truth only through "symbols and images," such as the world made in the form of the All to aid our remembrance.[97] So if ours is an image of true flesh, it can be like the rest of matter. It can be the way to knowledge, a means by which truth is known—the image through which we are reborn, because that's what knowing is. The Truth, leaving us no longer without truth, leaves ours the true flesh, and no mere image. We understand that it is part of the world in the image of the divinity that permeates it as its form.

Despite the vigor of the heresiologists, it seems clear that Valentinian texts are devout interpretations of Christian scripture. "Nothing is more certain than that the author of this treatise was attempting to think like Paul," writes M. J. Edwards of the *Epistle*.[98] "The resurrection," Rheginus's correspondent insists, ". . . is the truth which stands firm. It is the revelation of what is, and the transformation of things, and a transition into newness. For imperishability descends upon the perishable; the light flows down upon the darkness, swallowing it up; and the Pleroma fills up the deficiency. These are the symbols and the images of the resurrection. He it is who makes the good."[99] The heresiologists' objections miss the mark: a present re-arising is a transfiguration, not a denial, of the flesh; it is a newness revealed in the same old world. It is a resurrection because it is a renewal, not a single but a constant new beginning. Of course, from Tertullian's or Irenaeus's perspective, such a thoroughgoing theosis is no less heretical than the denial of a bodily resurrection. But for those who know the truth of this mystery, resurrection is revelation of the divinity of the world's flesh.

Curious Forms 2: Multiple Incarnations

Everything causes a scintillation.
  Christopher Toth, physicist, cited by Robert Macfarlane in *Underland*

Within the Abrahamic traditions, the most familiar version of resurrection is that of a single ensouled body, rising a single time. We have just seen one alternative, in which resurrection is a present epiphany of lived divinity throughout the world of matter. In another, we find not one but multiple resurrections. Here bodily perfection is still attained, just more gradually than in a single rising. There are several variations on the claim for multiple risings, but none will leave us with a sense that the material of the world is disanimate stuff that stays neatly in its place.

Serial resurrection is not quite the same as reincarnation the way we usually understand it—an individual soul occupying a temporally successive series of materially distinct bodies. Our first option involves a mutual body-and-soul transformation. Origen, for instance, states firmly, "Our teaching on the subject of the resurrection is not . . . derived from anything that we have heard on the doctrine of metempsychosis; but we know that the soul, which is immaterial and invisible in its nature, exists in no material place, without having a body suited to the nature of that place."[100] He is careful to distance himself both from the Stoics, for whom the cosmos as a whole is made again, and from the Pythagoreans, who—at least according to Porphyry—held that upon the death of one body the soul went on to be born into another.[101] Instead, body and soul are reborn, or continuously born, together, in ways suited to their cosmic position. The souls need changed bodies because souls are purified by the suffering they undergo in each life. The more pure the soul is, the higher in the heavens it rightly dwells. The body must change along with it, because not all bodies are suited to all locations.

In birth as we usually understand it, Origen says, the soul "casts off the integuments which it needed in the womb; and before doing this, it puts on another body suited for its life upon earth."[102] We usually think of this as a singular event. Origen argues instead that such discarding and acquiring of bodies occurs at every stage of the soul's move toward perfection in a perfect body. Like the changes in the body at birth, a new incarnation need not leave the matter of the earlier body behind altogether. Instead, the "new" body is a transformation of the old, driven by the transformation of the soul. Origen uses an image, drawn from Paul, of a seed that becomes an unrecognizably different plant that is still, clearly, the plant of *that* seed.[103] An acorn does not grow into a cornstalk. An oak is the product of its acorn, not the product now of one acorn and later of another. But Origen suggests a range of possibilities in interpreting the image. Sometimes, the soul may need to exchange material for other material. Other times, material is transformed so that it accords with the soul. Finally, new matter may be acquired to supplement the body that the soul already has. "Accordingly, [the soul] at one time puts off one body which was necessary before, but which is no longer adequate in its changed state, and it exchanges it for a second; and at another time it assumes another in addition to the former, which is needed as a better covering, suited to the purer ethereal regions of heaven."[104] The "sameness" of the body may include, it seems, something like the shedding of a seed's husk, the change of the body's form (perhaps discarding some matter in exchange for other matter), and the taking in of new stuff.

Origen understands the animate body's gradual, literal elevation as another teaching, albeit a rather subtle and hidden one, from Paul. Paul declares that there are different forms of glory for heavenly and earthly bodies, for the sun, moon, and stars, and among the stars. He adds, in an imperfectly informative parallel, "So it is with the resurrection of the dead."[105] On Origen's view, as soul and body are transfigured and perfected together, the ensouled body ultimately takes the form of a luminous celestial sphere—perfect in beauty and perfectly suited to this literally highest realm.[106] Because the whole universe is restored, all of its matter must be taken to this place. It seems possible, at least, that the matter from any husk discarded by one body might be taken up as a necessary addition by another, until all matter is perfected into the appropriate form of perfected souls—it is clear that an ensouled body has no need to be in a human-like form. The stellar perfected body is the ultimate stage of all flesh, and the perfect goodness of the soul is the perfect beauty of the body, making it a perfect adornment of the cosmos as both cosmic and cosmetic, so that the world itself is beauty. A cosmos in which some matter lay discarded and unperfected below the glory of the stars would remain imperfect. Through apokatastasis the world is returned to its origin, gathered in Wisdom.[107]

A similar picture of perfection appears later in the work of Nachmanides.[108] For him, too, all souls end up in a perfect state of celestial embodiment, getting there through a series of resurrections. Once more, successive incarnations allow body and soul to be perfected together. At each incarnation, the body joined to the soul becomes increasingly refined and good. It ends as a luminous heavenly body, a body of rarified but still material stuff.

Despite these striking similarities, though, there are substantial differences between Nachmanides's and Origen's accounts of successive incarnations. Part of the distinction has to do with the means of improvement—suffering that purifies versus action that mends.[109] In more narrowly somatic terms, Nachmanides does not seem to be especially concerned with the question of whether the matter of each successive body is conserved. He is more concerned with the way that the joint transformation of soul and flesh approaches the unsayable secret of divinity. "There is in this form deep secrets," he writes of the resurrected body, ". . . for the existence of the body will be like the existence of the soul, and the existence of the soul will be united with supernal knowledge."[110] Though this passage is elusive, perhaps the body itself knows the deep secret that is divine, and the way that a body knows is a secret beyond the intellect. Knowledge is transformative. From one incarnation to the next, the soul retains traces of memories it has gathered through successive bodily forms, helping it to come ever closer to the understanding that will in turn enable the action that heals the broken world. Nachmanides's remark on secrets suggests that some memory may be somatic. Glossing this passage, Elliot Wolfson writes,

> The cleaving to the supernal knowledge . . . is depicted as an augmented luminosity of the face and as being garbed in the Holy Spirit, characteristics that are adduced from several biblical and rabbinic figures. We may conclude, therefore, that for Nachmanides, the secret entails a somatic transformation that is in inverse relation to the mystery of incarnation, the glory assuming tangible shape when it appears in the world.[111]

The body as it becomes perfect approaches the embodiment of the mystery of divinity, flesh moving toward divinity as divinity moves into flesh. The mending of bodies unites them with mended soul and divine knowing as they restore the cosmos. Approaching perfection, they become less, not more, distinct from one another.

Wolfson explains that for Nachmanides, "The eschatological body will be like the original embodiment of Adam who, prior to the sin, was created to be immortal."[112] The cosmic Adam maps perfectly the beautiful cosmos, full of stars. Once more, the boundaries and distinctions of macro and microcosms refuse to lie tidily in place. Body is given a bit more credit than it was by Origen, for whom it is essential to perfection, but dependent upon soul for its form. But the results are similar in some important ways: when the body is finally regathered, each aspect of it is perfectly beautiful, perfectly integrated with soul—and its beauty is nothing other than the beauty of the cosmos.

Nachmanides's considerations are further developed by later Kabbalists. The *Zohar*, roughly contemporaneous with Nachmanides's writing, explains, "As long as a person is unsuccessful in his purpose in this world, the Holy One, blessed be He, uproots him and replants him over and over again."[113] Isaac Luria once more presents a particularly interesting reading of this idea. (It is also particularly complex, and I will draw out only one thread from the intricate weave of his discussions of resurrection.)

Luria's version shares with earlier Kabbalah the description of the fall of Adam as the breaking and scattering of his soul and body alike, mirrored in and mirroring the fragmentation of creation.[114] He draws on a reading of Genesis in which the first Adam includes the bodies and the souls of all subsequent humanity. "When God created humankind, he made them in the likeness of God," we read in the fifth book of Genesis. "Male and female he created them, and he blessed them and named them 'Humankind' ['Adam'] when they were created."[115] Though this is usually read as indicating that humanity will *descend from* Adam, it can also be read as the claim that all the matter and souls of humankind are *gathered in* Adam. An earlier midrash suggests something like this: "Even when the first man was a lifeless mass, the Holy One, blessed be He, showed him all the righteous men who would descend from him. Some hung from his head, others were suspended from his hair, and still others from his neck, his two eyes, his nose, his mouth, his ears, and his arms."[116] When Adam erred, his body was scattered into lesser human bodies. His soul, in which all human souls were collected, was scattered as all human souls into those bodies.[117] And the light of the divine source, which was so close to being regathered, scattered again throughout creation.[118]

That subsequent humans have to live in this confused and imperfect world is just, but justice without mercy destroys worlds. The two must be rejoined to restore a creation that has been broken. In regathering, the intent is not to wipe out evil, but to integrate one and two—"good impulse and evil impulse, interwoven, one," as one of the *Zohar*'s rabbis puts it. Another adds, "The wicked cause a defect above. What defect? The left is not integrated within right, for evil impulse is not integrated in good impulse because of the sins of humanity."[119] So sin causes the disintegration, but wickedness too will be reintegrated, preserving multiplicity in its unity. The breaking up and then the gradual reassemblage and perfection of the soul are described in terms of the slow reassembly of the body as well, though the body is "only" a "garment" for the soul.[120] The Adam who is dispersed into all bodies and souls is regathered into the perfect ensouled cosmic body.[121]

Though body may be less worthy than soul, it is still caught up in the complex and layered processes of resurrection. Luria does not speak in terms of earth as co-creator but does say that the human being is created of heaven and earth both, mirroring the creation of the cosmos in the first lines of Genesis. Like both Origen and Nachmanides, he emphasizes the point that each embodiment is a chance to perfect new levels of the soul.[122] But he recognizes another issue that arises if we do not, like Origen, insist that each successive body is somehow the same one. This is not unlike a problem with which the disciples confronted Jesus. Which marital partner, they demanded, will a person join in the afterlife?[123] And here we easily imagine the skeptic scoffing: which in its series of bodies is the finally perfected soul going to choose?

Luria, delightfully, answers, all of them. We read in the *Gate of Resurrections*, recorded by Chaim Vital, "When the time for resurrection arrives, each body will take its portion of soul according to the level [that was rectified] in its time."[124] What has once been mended or restored has no need for reembodiment. Besides, it would be wrong to split apart the soul-body that has perfected itself together.[125] The soul that is embodied begins as a root soul—the portion of the soul that first split from Adam's or was a shattered portion of it. As much of that soul as is perfected in one life remains with its first body. What has yet to be made right is sent out as a spark into a new embodiment, and so on again. The soul is distributed into successive bodies—or in certain cases, into a few bodies at once, if the root sends out several simultaneous sparks. Like the root, each spark remains with the body in which it was perfected.[126] The soul that remains with each body will in turn re-enliven and transform that body's remains at resurrection, awakening the properties of the luz bone so that it raises the body from dust like yeast raises dough. The collective restoration plays its part in the restoration of the world; the restoration of a collection of sparks and their root works together to restore the relevant limb of Adam's body, mirroring the limb of the cosmic body from which it was broken, until that body is made whole.

Each of these divergent processes of reincarnation is a process of restoration and of gathering. The bodies of the human, the first Adam, the cosmic Adam, and the cosmos itself gather slowly together into a perfect configuration. The body once more confuses itself with and within the cosmos. Matter, we must begin once more to suspect, exceeds its formations, bringing its own power into the final form of the perfect world. But the strangest roles for matter actually appear in the most conventional versions of bodily resurrection.

Curiouser and Curiouser: Regathered Matter

> For I have already been born as a boy and a girl, and a bush and a bird and a ⟨mute⟩ fish ⟨from the sea⟩.
>
> Empedocles, Fragment B117

In fact, the most orthodox versions of resurrection may turn out to be the most surprisingly strange overall. In theories of singular resurrections, the question of how a body remains the same one, despite the degree to which bodies vary over the course of a life, receives particularly detailed attention. The answers can be rather baroque.

The puzzle of how to regather bodies that might have dissolved into their elements or been eaten by other animals was already a popular one in the second century. In fact, Athenagoras is already impatient with it. "To disbelieve things which are not deserving of disbelief is the act of men who do not employ a sound judgment about the truth,"[127] he declares, and the claim that God would be able to regather the material of the body is a thing unworthy of disbelief. (In less convoluted terms, it is worth believing that God can regather the material of bodies.) As Athenagoras explains, two things make an act impossible: if a person does not know how it is done, and if that person does not have the ability to do it. God is not ignorant of any means, nor is any act beyond God's ability. So it is absurd to suggest that God is unable to re-collect bodies—we may also recall here Athenagoras's argument that this is an act that in any case is surely easier than what God has already done in creating those bodies in the first place.[128] In declaring a bodily resurrection, Athenagoras also makes a declaration about numbering. The body that dies must be materially and numerically the same as the body that is resurrected, because otherwise the concept of "resurrection" doesn't make sense. He draws on the Aristotelian notion of a final cause, the end that is a thing's aim or goal:

> There must by all means be a resurrection of the bodies which are dead, or even entirely dissolved, and the same men must be formed anew, since the law of nature ordains the end not absolutely, nor as the end of any men whatsoever, but of the same men who passed through the previous life; but it is impossible for the same men to be reconstituted unless the same bodies are restored to the same souls. But that the same soul should obtain the same body is impossible in any other way, and possible only by the resurrection.[129]

Athenagoras would probably be horrified by Isaac Luria's multiplying embodiments; for him resurrection must preserve personal and numerical identity, singular in body and soul, together. Likewise, Gregory of Nyssa writes, "We assert that the same body again as before, composed of the same atoms, is compacted around the soul" as part of the universal restoration. (We should note that in this, his version of restoration differs from Origen's, in which the matter of various bodies could be gathered together, newly distributed and configured, in the celestial spheres). Although "all are agreed that these bodies which the soul resumes derive their substance from the atoms of the universe," he writes, "they part company from us in thinking that they are not made out of identically the same atoms as those which in this mortal life grew around the soul."[130] The same atoms are gathered in a newly perfect form. Identity seems to be material rather than formal.

By the time that Thomas Aquinas writes, the Fourth Lateran Council of 1215 has affirmed a more detailed dogma of bodily resurrection. Once more judgment is central. "All . . . will rise with their own bodies, which they now wear," the Council declares, "so as to receive according to their deserts, whether these be good or bad."[131] Thomas agrees, but he finds other reasons as well to declare the resurrection of identical bodies. Like Athenagoras, he draws on Aristotle. He argues that what is part of the nature of a species must be found in all of its members. (If it is in the nature of fish to be aquatic, then all fish must be aquatic, or they aren't fish.) As Aristotle points out in *De Anima*, body and soul are a unit.[132] And if it is part of human nature to have or to be a body, all humans *must* have bodies; otherwise, they are incomplete and imperfect. "The soul cannot have the final perfection of the human species," Thomas writes, "so long as it is separated from the body. Hence no soul will remain forever separated from the body. Therefore it is necessary for all, as well as for one, to rise again."[133] The universe, because it is the creation of a perfect God, can only aim toward perfection. This demands the restoration of every soul-body unit, with the soul acting as the form of the bodily matter.[134]

Thomas, like both Athenagoras and Gregory, insists that the nature of resurrection demands that the same matter that disintegrates must be regathered.[135] Once risen with this numerical sameness, the matter—nonetheless the same—is not subject to any subsequent disintegration; "They will rise, then, never again to die."[136] All of our authors who posit a single, final resurrection agree upon this point, which Paul also emphasized strongly: risen flesh is not subject to corruption.

Augustine is among the first to think carefully through the implications of this identity, and he does one of the most thorough jobs of it. Augustine works hard to keep the idea of bodily resurrection reasonable. But its opponents, he notes, raise all sorts of questions about it, meant to show that the very notion is absurd or even impossible. He feels himself obligated to answer. Will miscarried infants rise? Augustine admits his own uncertainty here, but their rising seems likely to him.[137] How can bodies be without flaw or blemish and still be those that a person had in life—especially since few people die when their bodies are at their best? Relatedly, how "old" is the risen body; that is, what age of the person is recreated in risen flesh? Augustine makes Christ the model for subsequent resurrections: he assures his readers that people will rise with their bodies as they were at about the age of thirty, the age at which Christ died and rose. This is also his answer to another question: devout Christians say that all risen bodies will have the stature of Christ. But does this mean that some tall people are going to be shortened? (There is less concern with the potential stretching of shorter people; tallness seems to be associated with beauty, though it is unlikely that Jesus was tall. It is also impossible that people should lose any of their material, so the tall cannot be shrunk.) No, Augustine says; it means that we shall all rise as we were or would have been at "the bloom of youth," thus achieving "not the measure of the stature, but . . . the measure of the age of the fullness of Christ."[138] This applies even to infants, who had in potential the greater age and larger bodies that they never attained in life.[139] (Thomas Aquinas later agrees, assuring the reader that God can make up the missing matter from some other source.)[140] The body is resurrected as the best version of itself. Incidental blemishes will be removed, but some things that are not really flaws—such as the scars of martyrs—will remain as signs of glory. For all these changes, the body is, emphatically, materially what it was. The exception is that those who died in infancy may have matter added in from elsewhere, at least according to Thomas's interpretation of Augustine. It is rather fascinating that these defenders of creation *ex nihilo* (including Athenagoras as well) assume that the matter of resurrection must preexist the resurrected state, rather than assuming that God might make more matter as needed, as for instance to fill out the infant form.

The regathering of a body's matter presents some of the most interesting issues in Augustine's discussion. He agrees that all of the matter of the body must be regathered. But he can easily imagine another skeptical inquiry: won't we look strange, with all of the length of our trimmed toenails returning, with all of the hairs we have cut or shed reattached? Returning once more to the dust, Augustine uses the familiar metaphor of molded clay for the matter of the risen body. All of the "clay" will be gathered together, but it will be remodeled into a new form. That form will be the most beautiful possible; that is, the best proportioned version of itself.[141]

Augustine recognizes further problems with bodies that have not only disintegrated but have been widely scattered. He assures us that "all the portions which have been consumed by beasts or fire, or have been dissolved into dust or ashes, or have decomposed into water, or evaporated into the air," will be gathered back into the clay to be remolded. In fact, he does not think that this is especially difficult.[142] Flesh may be regathered, even if it has been scattered; it may be redistributed in forming an individual, so that it assumes a particularly pleasing form. Scattering is essentially just decomposition over a wider space. Redistribution will present a greater problem, but we might best make sense of it by moving forward a bit, to draw more ideas into the conversation.

Saadia asks some of the same old questions that had troubled Augustine, but with added complications.

> One might then ponder: when the first generation of humans died, their elements separated and each part returned to its source—heat returned to fire and mixed with it, and humidity to air, and cold to water, and dryness remained earth—and the Creator then composed the bodies of the second generation, which already contained in them parts of the first. Then the second generation of human beings died

and their parts also mixed with the sources of the elements, and the Creator again composed of them a third generation whose end was like that of the two previous ones, and so with the fourth and the fifth. How, then, will it be possible to return the first generation in its entirety, and the second in its entirety, and the third in its entirety, when the whole of the second generation is already slightly mixed in?[143]

Saadia's answer is simpler and tidier than Augustine's (or Thomas's after him). Believing as he does in the resurrection of all bodies at once, on this earth in their material state, he affirms that sufficient material and space exist to allow unmingled matter for all bodies.[144] Thus the first possessor of the bodily matter is the last and only one.

We have worked our way to one of the most challenging problems of same-body resurrection: some bodies may be consumed by others. What, Augustine frets, becomes of "the flesh of a dead man, which has become the flesh of a living man?"[145] Earlier writers, including Athenagoras and Tertullian, had already assured their readers that human flesh devoured by *other* animals would be returned to the human beings, as rightly belonging to the "superior" species.[146] Athenagoras pushes on to the question of humans consumed by other humans, whether directly ("the devouring of offspring perpetrated by people in famine and madness, and the children eaten by their own parents through the contrivance of enemies") or indirectly, when animals who have consumed human flesh are later eaten by other humans.[147] He rather ingeniously determines that human bodies are not properly digestible, and so never do form parts of other bodies. This simplifies the divine act of regathering human flesh.[148]

Augustine grants that cannibalism is rare, but he does not assume the indigestibility of humans. He determines that the eaten flesh "shall be restored to the man in whom it first became human flesh. For it must be looked upon as borrowed by the other person, and, like a pecuniary loan, must be returned to the lender."[149] Thomas will largely agree: "Therefore, the flesh consumed will rise in him in whom it was first perfected by the rational soul."[150] (There is room here for a loophole, and we shall soon have to look into it.) Augustine adds that a cannibal who was driven to consume human flesh by famine need not fear being raised emaciated in consequence of the loss of that flesh. As with infants, God will fill out the ideal possibilities of that body so that it is maximally beautiful.[151] The matter will just be drawn from somewhere else.

Like Augustine, and modeling his answer in part upon *City of God*, Thomas Aquinas also agrees that all parts of the body shall be regathered and remolded, including a lifetime's worth of hair and nails, with all the body's loss and damage repaired.¹⁵² But Thomas, with his more complicated Aristotelian categories, marks exclusions to this regathering. These include sweat and urine and milk and semen. The former two, he says, are excreted from the body because they make it imperfect. The body will not be resurrected with its imperfections, as they are not part of it in the most proper sense—that is, they make it an imperfect instance of itself. The latter pair will not be resurrected because they do not belong properly to the individual from whom they flow. Rather, they contribute to the general preservation of the species.¹⁵³ A species as such is not resurrected; only its members are. Thomas is determined to maintain the individual coherence of bodies even when life has not done so, and even when that coherence is threatened by bodily exchange or by the inextricability of the "individual" from the "species." This reasonable insistence on the individual falters, though, when we face an especially strange redistribution of matter.

This strangest redistribution occurs when species preservation intersects with cannibalism.¹⁵⁴ We've seen that when animals eat human flesh, humans get it back. And in individual instances of cannibalism, Thomas follows Augustine in returning the flesh to the one who had it first, with God taking some of the world's other matter to fill in the otherwise too-thin body of the cannibal. But cannibalism can have more complicated consequences. Thomas summarizes the problem: "It happens, occasionally, that some men feed on human flesh, and they are nourished on this nutriment only, and those so nourished generate sons." (We might be given pause by the breezy assertion that "it happens," but let us continue.) The semen with which they generate those sons, or the milk with which they feed those sons, is properly part of the sons' flesh and not of their own. (We have to assume here, I think, that Thomas includes mothers and daughters, unless he has in mind contributors of semen who can also gestate and lactate.) "Therefore, the same flesh is found in many men. But it is not possible that it should arise in many."¹⁵⁵ The problem is particularly acute if the parents have eaten nothing but human flesh, so that the entire substance of their seed and their milk must itself be nourished by—made of—other people's matter. Thomas says that the seed, even though it is made of someone else's matter, rises with the person who was generated from it—otherwise, that person would have no being at all. This accords with Thomas's general idea that semen arises with the one generated by it and not the one from whom it comes, because it is transformed into part of the generated person. (We can return to Paul's metaphors to make sense of this: if a botanical seed were resurrected, it would be as the plant that grew from it, not the plant from which it came.) As for the person eaten, "there will be supplied in him . . . something from another source. For in the resurrection this situation will obtain: If something was materially present in many men, it will rise in him to whose perfection it belonged more intimately."[156]

This last point, like the remark that flesh belongs to "him in whom it was first perfected by the rational soul," marks an important difference from our previous answers. Usually flesh arises as part of its first formation, Thomas agrees. But when it been molded by many forms, it will retain the most perfect. Because the cosmos intends perfection, when matter cannot be clearly sorted out as belonging to one body, that matter goes toward its most perfect form; that is, to the best of the people who had it. In the case of equal perfection, the matter goes to the one who had it first.[157] It could be that souls take up matter. But it could be that matter here has its own agency; in a strange sense, it chooses its soul. If any matter that is needed for a perfect form is missing, that matter is made up from a vague "other source," once more taking in more of the world.

This complexity requires summation, though it cannot quite be gathered together neatly. Athenagoras and Saadia give relatively straightforward answers to the problem of shared matter by making sharing illusory in some way, with indigestible humans or with matter enough to make sharing unnecessary. Augustine and Thomas Aquinas, agreeing that bodies rise numerically the same, both attempt mightily to sort out the matter of bodies: a body is not the same body if it is not made from the same stuff, even if some qualities of that stuff must be transfigured or its order and configuration improved. But both falter in the face of quantity. They can almost cope with too much stuff, by enlarging the scale of the resurrected. But too little stuff and shared stuff present a problem of deficiency: something must be made up from somewhere. They do not simply assume that God would make more matter to tidy up and even out the distribution. Even here, however unintentionally, these thinkers seem to hold a lingering respect for the creative and primal role of materiality. They meet Nachmanides and the earlier rabbis, for whom the dust was worth God's consultation. And where no other factor sorts out that dust, matter sorts itself to join the best of its series of forms.

In every one of these forms of resurrection—immanent, successive, and singular—something important happens. The flesh we each tend to think of as "mine," and to assume will eventually be "mine" again, eludes such tidy possession. Immanent resurrection is a revelation that is given to those who know the whole world as the image of divinity and themselves as members of that image. As we saw in chapter 3, that image is as complete in every point as in every other. The story of flesh that is successively transfigured, as Origen thought, or the story of successive bodies that are resurrected in perfection for having perfected the parts of a soul, as Nachmanides and Isaac Luria held, undoes the very possibility of numerical individuation, one set of atoms matched forever to one soul. In consequence, the cosmos itself becomes both material and animate. Finally, the insistence on the same atoms re-arising with the same soul founders against those atoms' dizzying collection and redistribution. Despite more than a millennium of the best efforts of these quite brilliant theological minds, bodies' tendencies to interchange and overflow present persistent problems, more easily opened than resolved. The matter of the body is too much for a single and self-contained incarnation of the soul. If it moves toward a new heaven, it is too shared for a single full resurrection. If it returns to a first creation, it has strangely gathered beauty along the way, which it nonetheless brings back to its origin—an origin in wisdom that is beauty itself.

The Body's Beauty Lives

> Beauty is momentary in the mind—
> The fitful tracing of a portal;
> But in the flesh it is immortal.
>
> The body dies; the body's beauty lives.
>
> Wallace Stevens, "Peter Quince at the Clavier"

Finally, we begin again from this chapter's starting point—resurrection on the scale of the cosmos, a recurrence of every thing and every event, a world that is destroyed and recreated. Unlike apokatastasis, eternal recurrence does not stop after reaching back to its starting point but repeats the entire process. Though the idea has never been especially mainstream, it does itself appear again and again. Origen connects it both to the Pythagoreans and to "the learned among the Egyptians."[158] But the idea is most strongly associated with the Stoics, who called the rebeginning *ekpyrosis*, the dissolution of the god-matter of the world into cosmic fire, from which it reemerges always the same again.[159]

Stoic thinkers varied in their certainty about recurrence, but by the fourth century the Christian bishop Nemesius could describe *ekpyrosis* as Stoic doctrine: "The periodic return of everything occurs not once but many times; or rather, the same things return infinitely and without end. . . . There will be nothing strange in comparison with what occurred previously, but everything will be the same with no difference down to the smallest details."[160] In the late nineteenth century, Friedrich Nietzsche, whom we earlier saw delighting in the Pythagorean physicality of poetry and rhythm, seizes upon the idea of recurrence and brings it back to a broader philosophical awareness. He offers a description strikingly like Nemesius's: "This life as you now live it and have lived it, you will have to live once more and innumerable times more; and there will be nothing new in it, but every pain and every joy and every thought and sigh and everything unutterably small or great in your life will have to return to you, and in the same succession and sequence."[161]

Nietzsche draws on and credits the Stoics for this idea, but he does not adopt their notion of a cycle that is divinely rational. Instead, he also draws upon the pre-Socratic philosopher Empedocles, for whom things in the world are never stable, "but in that they never cease interchanging continually, in this way they are always unchanging in a cycle."[162] Nietzsche's version of the cycle, in the cosmos "as a play of forces and waves of forces, at the same time one and many," is "eternally changing, eternally coming back." So long as parts or forces are not infinite in quantity, this coming back leads to the infinite recurrence of particular formations, and then of particular series of those formations. Eventually, every part must interact with every other in every possible way, making of the cosmos a recurrent "mystery world."[163] In this mysterious return, beginning itself becomes untidy—it could be anywhere in the continuous cycle. Nietzsche's understanding of the spatial interchange of moving forces, like Empedocles's spatial interchange of moving elements, leads him to a strange temporal understanding as well. It is not just that beginning is unmarkable. Return itself becomes something other than a simple circle of what is identical. More exactly, "the same each time" and "not even the same as itself" emerge from the same revelation. Though our trio of resurrection stories seems very distant from eternal recurrence, all four share a mad interchange, whether of particles or forces or animae. The sharing disrupts a neat sense of well-bounded individuality. Given this disruption, it is less surprising that we can think the "individual" and the "cosmic" resurrections together.

The trio also share an emphasis on the beauty of risen bodies, however they may be understood. Beauty turns out to have a special relation to eternal return as well. For Nietzsche, it is the very source of the idea. In *Ecce Homo*, Nietzsche describes the thought of eternal return as "the highest formula of a Yea-saying to life that can ever be attained."[164] His own exuberant affirmation, the one that led him to the concept of recurrence as something more than a philosophical abstraction, was inspired by beauty. In a letter to his friend Peter Gast, he cites both music and mountainous landscape as sources.[165] The sheer delight of such beauty leads him to wish, or more insistently to will, that the perfect, beautiful moment remain forever. But this won't work; a frozen moment loses some of its beauty—especially in music. So what Nietzsche finds himself saying "yes" to is not endurance, but the infinite return of that moment. It cannot stay, but it can come around again and again. With that yes, he experiences, just for that moment, what a return would be like.

One of his twentieth-century interpreters considers the implications of pushing such assent to its limits. Pierre Klossowski (who was both a translator of Augustine and a student of Gnostic texts) notes the epiphanic character of the thought of return, which he calls a "lived fact" and a "sudden thought."[166] It is, he says, revelatory before and even when it is propositional.[167] The revelation only gradually brings propositions with it, as one works through the suddenness afterward. Nietzsche had already argued that to affirm a moment—to have the sudden lived sensation of its recurrence—must be to affirm the life that extends before and after it, making it the moment that it is.[168] Klossowski expands this idea to consider the revelatory moment in the context, beyond a single life, of a cosmic recurrence. Here a full affirmation of a moment would mean that every possible way of leading up to it, and every possible afterward leading from it, would also have to be affirmed. Every interchange would have to happen, in every order. From one lifetime, the affirmation of return would have to extend to uncountably many possible lives. Thus, says Klossowski, "The revelation of the Eternal Return necessarily brings on the successive realizations of all possible identities. 'All the names of history finally are me.'"[169] But in a kind of double movement that we might now find almost familiar, Klossowski also realizes that most versions of the cosmos will not include a "me" at all. "At the moment when I am struck by the sudden revelation of the Eternal Return," he writes, "*I no longer am*."[170] To affirm everything is a scattering and reordering of everything, in which one both finds and loses oneself in everything everywhen—a scattering and recollection that we can recognize from other modes of resurrection. And this thought is brought on by beauty, in all its complication.

Even though the recurrence of the cosmos and the endless interchange of everything in it could be forms of resurrection, the connection of bodily resurrection to Klossowski's epiphanic reading may seem a bit forced. But a remarkably similar affirmation, in the form of maximum delight taken in incomprehensible beauty, also marks descriptions of bodily resurrection.

This oddness brings together the beauty of bodies with Paul's strange insistence on their incorruptibility. None of our sources suggests that incorruption means stasis. Instead, we read of movement, of sensations, of songs of praise. Singular glorious resurrections turn out to be multiple in their impossible gathering and division of material creation, all selves and none drifting in the world's holy dust. Through successive (or simultaneous) multiple embodiments, embodied forms come not to deathly stillness but back to their starts, luminous and sparkling even when still. And in an immanent resurrection, the eye-twinkle of a moment pulls time out of time, makes resurrection precede death, makes living now into having risen. Klossowski's reading of maximum affirmation can help us theorize the way that matter is scattered and gathered and the ways that beauty is more than an incidental property.

The beauty of resurrected bodies is always a bit strange. It may be a reward, a revelation, or even a necessary aspect of perfection. In what may be the most commonly shared characteristic among descriptions of risen beauty, the resurrected body is often described as luminous. The star-like perfected bodies, says Gregory of Nyssa, have "a brighter and more entrancing beauty" than we perceive now.[171] For Origen, perfected matter "shines in the splendour of 'celestial bodies,'"[172] and the saints have "bodies . . . of a shining light."[173] In theories of single resurrections, Saadia writes that the light of Wisdom makes the final bodies of souls even more brilliant than the shining stars.[174] Thomas Aquinas offers a long list of distinctions between risen and condemned bodies: lightweight or burdensome, obedient or resistant to the soul's commands, "dense and darksome" or "lightsome," capable or incapable of suffering, fulfilled or hopelessly longing.[175] The Talmud declares that "the radiance of the righteous in the future will be three hundred and forty-three times more luminous than the radiance of the sun when it goes forth in its light."[176] Both canonical and pseudepigraphal scriptures offer descriptions of resurrected bodies as beautifully luminous.[177]

Luminosity makes an interesting figure of beauty because it allows beauty to perform a revelatory function. Light may be somehow beautiful in its own right, but is also what makes anything else visible. We are returned to the puzzles of the self-seeing eye, or of Adam as the light revealing itself. Luminous beauty reveals beauty. When some of our writers describe beauty as a reward for goodness, it is a reward that simply reveals what is true. Beauty makes goodness visible. Gregory of Nyssa argues that the beauty of the risen form reflects the virtue of one formed. The more virtuous the person, the more beautiful the resurrected body.[178] The pseudepigraphic text 2 Baruch reiterates the idea of beauty and ugliness as reward and punishment but adds that the transformation takes in perception along with appearance: "For they shall behold the world which is now invisible to them,/And they shall behold the time which is now hidden from them:/And time shall no longer age them."[179] The beautiful perceive the wondrousness that is currently invisible because its time is hidden. The transformation of the person is mutually the revelation of the wonderful. Both are beautiful.

Thus the rewarding increase in beauty is also a matter of proper perception, bringing other understandings of resurrection closer to revelatory gnosis. To perceive rightly, perhaps by the light that beauty is, may be its own reward or part of the reward of resurrection. Origen describes a transformation that occurs all along the way to the final restoration, and in it he connects beauty to re-arising. He urges his readers to "walk in newness of life, showing ourselves to be daily new and, I might say, more beautiful to him who raised us with Christ. Let us be transformed into the same image, collecting the beauty of our face in Christ as if in a mirror and observing in him the glory of the Lord, by which Christ, rising from the dead, ascended from earthly humility to the glory of the Father's majesty."[180] The beauty of the beholder becomes downright glorious as a mirror of the beheld, and the mortal ascends to divinity. The *Gospel of Philip* too takes up the mutuality of perceiver and perceived, but with the declaration that it applies only "in the realm of truth" and not in the world without epiphany. It belongs to those who have become one with the Son:

> People cannot see anything that really is without becoming like it. It is not so with people in the world, who see the sun without becoming like the sun and see the sky and earth and everything else without becoming them. Rather, in the realm of truth, you have seen things there and have become those things. . . . [Here] in the world you see everything but do not [see] yourself, but there in that realm you see yourself, and you will [become] what you see.[181]

That place is this place, properly seen. The revelation of beauty is the proposition of the world's divinity. Beauty comes together with newness to remind us of Wisdom, who works with the creator to perpetually delight in the novel creation of the world.

Delight cannot properly be disentangled from beauty. The perception of beauty by beauty, light by light, is itself delightful. Descriptions of risen beauty may include the delight that it provokes. For Augustine, as we have noted, martyrs' scars not only become acceptable, but are perceived as beautiful themselves as a part of the world's new beauty.[182] So too do the bodily markers of sex, unnecessary for reproduction but "adapted . . . to a new beauty, which . . . shall excite praise to the wisdom and clemency of God."[183] This excitation is not restricted to those markers; resurrected bodies are of such beauty that the very sight of them moves other risen persons to praise for the body's creator.[184] In this Augustine echoes the book of Enoch, where we read of the happily resurrected, "And their mouth shall be full of blessing. And their lips extol the name of the Lord of Spirits."[185]

"The body's beauty lives," says Wallace Stevens, and what is divine and delightful in the world is the life of it, its own divinity. It lives again whenever it is known. And it is known not as an intellectual deduction, but as a revelation through the delight that recognizes beauty and loves it. This revelation is itself transformative and calls forth amazement. The strongest affirmation expands the moment to all moments, as it must, since no moment occurs in isolation. Nietzsche might well have said of the return, "It is the revelation of what is, and the transformation of things, and a transition into newness." But Rheginus's correspondent had already said it.

Affirmation is most obvious in immanent resurrection but is not restricted to it. For believers in a singular bodily resurrection, the long processes of life on earth leave beings all caught up in one another's matter, at once all things and no singular thing or identity at all—the same revelation that accompanies an affirmation of truth and desire or of the beauty that demands a yes. For those who hold to a series of embodiments, matter's neat numbering is only overlaid on its transfigurability. Or soul, which forms, is split and resplit across a range of formed matter—which is, "finally," its gathering of all in all, into the instant that is infinite and vibrant—and is an origin that is also an end, leaving time strange. This is not return as the burden of existence, the dullness of over again, but the startle of the infinite within the finite, the immortal in the now. When we know the flesh as immortal, it is not as a fact, but as a delight, or perhaps an astonishment. This astonished delight is not containable by the bounds of an individual. Human singularity and superiority cannot hold up where the animate matter of the world is all so imperfectly distinguishable, nor where knowing the world requires a susceptibility to its beauty. Neither the time nor the place nor the stuff of the agent and knower holds on to itself. Astonishment rises anew.

Afterword

Of Love and Delight

> Some men say an army of horse and some men say an army on foot
> and some men say an army of ships is the most beautiful thing
> on the black earth. But I say it is what you love.
> 			Sappho, Fragment 16, translated by Anne Carson

Delight is no minor matter. Like judgment, it is a kind of assent. Judgment as assent is a central concept in Stoic epistemology, which influences both Middle and neo-Platonic thought after it, as well as having a more direct influence on thinkers including Augustine and later Descartes. In such judgment, we receive impressions and we have impulses toward them, including whether or not to think that they are accurate. But we assent to them only when we have judged them to be in accordance with reason. For Descartes, this reason is logic, but it is founded in our ability to recognize by the "power of understanding or . . . light of nature," which is a gift from God, meant to keep us from error.[1] The Stoic connection is more direct: reason is the divinity of nature, displayed in nature's laws and systems. Judgment's assent says that something is true in the sense of being a fact. Judgment is assent to reason, a yes to argumentation and rational speech. The assent of delight says that something is desirable—that it is beautiful, where beauty takes on the Platonic sense of desire-drawing.[2] Peter Brown points out that for Augustine, who knew Stoic epistemology and psychology well, delight is not only "the mainspring of human action," but in fact its "only possible source, [since] nothing else can move the will."[3] Intellectual assent may be more sedate than delight is, but both are responses that move us, and both are reactions to kinds of truth. One enables understanding, one transforms, and the two work together. Intellectual understanding can change the knower—for the Stoics, recognizing the world's divine rationality transforms every aspect of the sage's life. Transformative revelation can create new knowledge—for the Platonists, the ability to respond to beauty, which is love, leads to knowing the divinity of creation. Love, in turn, may recognize as beautiful what would be ordinary if unloved, revealing the ordinary as already divine.

Not all delight is mythic, of course, but its transformative element tells us something of the way that myth works. Not all myths are delightful, either. But Wisdom stories, where rejoicing is the very process of creation, are explicitly sources of delightfulness. The exegetical myths and commentaries created to think more about Wisdom share delight as well; it sneaks its way past ecclesiastical disapproval, past scowling punitive gods, past the accusations of a lifelong immaturity. "The gospel of truth is joy," begins the text by that name.[4] Though the joy is qualified as belonging to those who have received truth, it has already offered a therapy not altogether unlike Pythagorean dancing, a possibility that entices. The truth, after all, is a joy itself.

One of the best presentations of double assent appears in the *Gospel of Philip*, where Knowledge and (loving, joyful) Truth must work together for those seeking the world-revealing epiphany of resurrection. Though the text remains specific to its tradition, I think that it also offers a paradigmatic affirmation of the matter of the world, weaving together what we can learn through reason and myth.

People who have only knowledge, says the text, often become arrogant.[5] This is the very arrogance that Irenaeus is sure characterizes all gnosis. But really, they have only half of the necessary transfiguration: "Truth is the Mother, knowledge is the father."[6] The two are complements and consorts. As parents, the pair reproduce. Knowledge and Truth are passed on to another generation by teaching. (The Valentinians, we recall, are promiscuous and enthusiastic teachers.) Teaching requires knowledge; there must be something to teach. Teaching the knowledge of the elusive unknowable, though, requires Truth that loves. The strange whole-part relation returns: "Truth is one and many, for our sakes, to teach us about the one, in love, through the many."[7] Because of love, truth makes itself teachable. Those who have learned to love will go on to teach further—that is, to share their knowledge. It is here that we read, "Whoever is free through knowledge is a slave because of love for those who do not yet have freedom of knowledge."[8] Gnosis is not permission to go off on one's own to a glorious new freedom; instead, it imposes the requirement that others be loved and therefore taught.

This might give the impression that knowledge frees and love enslaves, but in fact love shares the freedom that knowledge gives. This is because love is not individually bounded: "Love [never says] it owns something, [though] it owns [everything]. Love does not [say, 'This is mine'] or 'That is mine,' but rather, '[All that is mine] is yours.'"[9] It is a kind of joyful generosity that makes sharing and dispossession something other than a burden. It is parallel to faith; love gives what faith receives.[10] And this brings us back to beauty. It is not especially laborious to share the love that is the revelation of Truth: "Spiritual love is wine and perfume. People who anoint themselves with it enjoy it, and while these people are present, others who are around also enjoy it."[11] The love of the world transforms that world into the truth of delight everywhere. And so we know truth in a double assent, both affective and propositional, factual and transfiguring.

Such stories tell us something of desire, too. For Plato, again, desirous love is the response to beauty.[12] For Aristotle, desire directs us to divinity; it is the force that moves anything toward its end or final cause, including the perfect cause that is divine.[13] Late ancient senses of the delight and desire go together so readily that *Desire and Delight* makes immediate sense as the title of Margaret Miles's influential study of Augustine's *Confessions*.[14] The Platonic version influences Sigmund Freud and all of psychoanalysis after him, where desire is a multidirectional force that seeks what it is missing.[15] But after the nineteenth century, despite the persistence of the Freudian version, we have tended especially to think of desire in Hegelian terms. Hegel, himself quite familiar with Christian theology, agrees that desire is the motivating force of human action. For him, however, desire negates, even annihilates—when we desire a thing, we will act to consume or subsume it, to make it part of us or to make it reflect us.[16] We desire to negate the things that are not us and thereby to expand ourselves further into the world. That desire might pause us, might even undo us, becomes a strange notion. But what might desire tell us, if not to take everything?

One possible answer, of course, is that it says, "[All that is mine] is yours." This is neither selfless nor selfish; instead, it is a complication of distinction. Such a desire—to share transformation—is at once "enslaved" and radiant, enough so to delight itself even as others breathe in its beauty. Wisdom's delight in the world tells us to rethink desire with the same attentive kindness that might want to share wine or perfume.

In *Gardens: An Essay on the Human Condition*, Robert Pogue Harrison offers a reading of the human relation to the garden of Paradise in Genesis that is at once creative and persuasive. In the second creation story, humans are instructed to care for and keep the earth, but the very ease of Edenic life makes them careless. Everything they might want is theirs immediately, theirs for the taking and the consuming (we might note one exception: the animals have been sent as companions, and are not eaten or, pre-exile, skinned. Neither, however, do they make much of an appearance once there are two humans). The humans' desire has no reason to be anything but negating; the world is at their disposal and for their consumption. Ease makes them arrogant, and they can see no reason not to eat any fruit they happen to want, even that offered by the sneaky serpent—the fruit of knowledge, we might add, valued more than the love that the garden embodies.

So banishment from this Paradise becomes something more and other than punishment. Exile offers the humans a real sense of caring as keeping, Harrison says. In the terms we have developed here, it allows them to desire to share themselves with the earth. The command that made no sense within the garden, the command to take care of and to keep well something other than themselves, becomes understandable and executable in the world. Harrison acknowledges that there is still a sense of accursedness here—"Care burdens us with many indignities," he notes. But we err in our "tendency to associate this putative curse with the earth, to see the earth as the matrix of pain, death, corruption, and tragedy rather than the matrix of life, growth, appearance, and form. It is no doubt a curse that we do not properly value what has been freely given as long as we are its daily beneficiaries."[17] I would add that if humans are earth themselves, creatures of clay or dust, they both devalue and misunderstand themselves in this view of earth as punitive.

Harrison turns to the beauties and benefits of cultivation—to its need for foresight, patience, and constant attention as the gardener transforms the landscape. Importantly, gardeners cultivate themselves along with their plots. (We might add that agency is shared among people and plants, sunlight and water, insects and dirt.) The desire to transform the world as a gardener might is certainly more benevolent and more widely beneficial than the all-consuming negative transformation that Hegel proposes. "All that is mine is yours" can be what love says not simply between people, but throughout the earth—and it becomes knowledge, too, as we realize the deep intermixing of all that is. The bounds that are me are a real part of my living and experiencing. But my world may also be known as strangely unbinding, scattering me with all the rest. Delight, that is, may oppose its desire both to destruction and to the imposition of a single self.

Of course, we can imagine delighting in something without loving it, perhaps because of the brevity of the experience; or loving what does not delight us, perhaps through the kind of care that is more purely a matter of duty. But the responses are hard to sort from one another in the context of Wisdom, where love, joy, delight, and revelation entangle. By delight we not only perceive beauty but help to create it. This is not an imposition of self onto world, but the self caught up in the worldwide movement of creativity. Wisdom that delights us undoes "us," decenters all that made us arrogant, returns us to the dust in all its astonishing capability. It does not diminish us but undoes the false diminution of the rest of the world.

The Loss of Mystery

> Most mornings I would be more or less insane,
> The newspapers would arrive with their careless stories,
> The news would pour out of various devices
> Interrupted by attempts to sell products to the unseen.
> . . .
> Slowly I would get to pen and paper,
> Make my poems for others unseen and unborn.
>
> Muriel Rukeyser, "Poem"

In myths of mystery and Wisdom, we think the world as new as a way to express the surprise of it. Careless impatience has no time for surprises, though, and the demand for use value has no time for patience. When we want answers, we become disinterested in raising more questions, the way that mysteries do. Then Knowledge discards loving Truth. Even myths can be read to tell us simply that we are in charge and should start making things over to suit our very important selves. If we really are trying to tell mysteries, we have to affirm that we are saying what we do not know. At a minimum, this opens the possibility that we might use the stories of anthropomorphic creatures to go beyond ourselves, beyond our agency and knowledge. We might use them as starting points for thinking about flesh, and we might find that flesh cannot at all be confined to bodies like ours; we start in ourselves, but myths are our ways of trying to tell ourselves more. (And after all, Merleau-Ponty begins his seminal discussion of the world's flesh by considering human sight and the use of language. We start where we are.)

Thus it is that the sheer arrogance of making the human the map of the cosmos turns out to be the same move by which gods and dust respect each other's creative agency, and the human is scattered and gathered among them, no more powerful over one than over the other. The redeemer who teaches humanity teaches that the entire world is as much in divinity's form as each person is, and that this form is never still enough to be known, never compliant enough to be overwhelmed by human action. The presumptuousness of thinking that humans overcome even mortality scatters and gathers the world's flesh with ours, in an act of assent that keeps us from containing ourselves. To think that materialism ends mystery is to forget how decentering mysticism has always been. This is as true of the self lost into the world as it is for the self taken out of it. Where divinity is diffused, the human self can hardly hold as a center.

Trying to think the world's delightful strangeness, I was led to Wisdom particularly by the sense of divine immanence, of a god-saturated world—and, to be sure, by her stubborn femininity persisting in often patriarchal traditions. But what I had not quite realized when I began was that Wisdom would be always beginning. She is co-creative from before the start, but her start does not stop; she keeps creation emergent and new. With Wisdom infusing the world, rather than making and forgetting it, creation becomes not a past but an ongoing event. And it comes to be not as something externally imposed, but as the creative force throughout the world. The narrative line of myth twists into an imperfect circle, where before and after, genesis and heaven reach toward one another without becoming the same. Anywhere on this imperfect circle can become a beginning, a place where Wisdom begins again.

This must seem to ignore something obvious, to move too quickly past carelessness and impatience. Delight's assent may be powerful, but it is much easier to forget than is judgment. It is easy to see dust only as what gets in the way of light, or as humanity's punishment for not being godly enough. It is easy to understand Limit in creation as a series of separations from anything divine. It is easy to read resurrection as clinging to oneself against all change forever. And it is easier than any of these to disregard myths; facts are far more useful in helping us to make the world the instrument by which we impose ourselves on all of the cosmos that we can reach.

Wisdom's newness is not the relentless pursuit of the next, but the continual revelation of what we had missed before, the same old, newly begun. And there is no small strangeness in speaking of beginnings now, when there is such a collective disregard for mystery in favor of utility and negation, to the point of ending many possible sources for the uprising of the new. Without wisdom, or when nothing can astonish us as new, death becomes more real than living, annihilation more real than radiance. Rushing selfishly and anthropocentrically to our own ending, we take much of the rest of the world's life with us. We legitimately wonder whether beginning might end.

"Each time my heart is broken," writes the poet Frank O'Hara, "it makes me feel more adventurous . . . but one of these days there'll be nothing left with which to venture forth."[18] This possibility of heartbreak emerges whenever we care. The temptation, of course, is to stop caring—to allow ourselves to be unwise, or to think, as I often find myself doing, that wisdom itself advises an end to these ventures. It is tempting to stop venturing forth while some part of our hearts remains unfractured. And it may be that all I have done in unfolding these exegeses of divine mystery from these ancient and unexpected sources is to write a long elegy for the tradition of telling a delight-filled world, an elegy to be deciphered inaccurately, after the humans are gone, by the descendants of resilient cockroaches and dexterous racoons. I hope that it may be more than that, that perhaps we can still attend, still pause to wonder. That wisdom will find adventurousness enough to go forth once more. That we may desire with the care that keeps and changes, with the delight that says "all this is yours" to the world where it must be true. With attention, we may at least hear the echoes and possibilities of an exuberant world. In this time, when endings become so much more evident, perhaps we can remember how to begin.

Acknowledgments

This book has benefitted even more than most from other people's insights and patience. My acknowledgments are almost certainly incomplete as well as inadequate, but no less genuine for that.

I owe thanks first to two remarkable groups of scholars. The members of LARCeNY, who explore late ancient religion in central New York, constantly model the combination of incision and generosity to which all scholarship should aspire. Among them, I owe particular thanks to Virginia Burrus, Jennifer Glancy, and Patricia Cox Miller for their conversation and kindness. I am indebted to Virginia, along with Mary-Jane Rubenstein, for gently showing me years ago how I was misconceiving pan(en)theism. Jennifer, who is unparalleled at giving others opportunities for their own thinking, also curated a lecture series that gave me a chance to consider the material on myth that is in chapter 1 here. I am indebted to Pat for many things, including the pleasures of our ongoing reading duet, where we choose books just because we like the sound of them. Most of the epigraphs here are taken from texts that we read together.

The second group is the 2016–17 set of fellows at Cornell University's Society for the Humanities, then under the direction of Tim Conway-Murray. The group's exceptional synergy combined with Cornell's resources (I still daydream about the library) allowed me to do a range of reading and thinking that I would never have managed otherwise. I am especially grateful to those with whom the exchange of ideas continues—Gemma Angel, Pamela Gilbert, Gloria Kim, Emily Rials, Elyse Semerdjian, and especially my Ithaca roommate Alicia Imperiale, who was sometimes able to get my very temporal brain to think about spaces.

Other conversation partners have been valuable. Gary Mar gave me more fascinating information about fractals than I have been able to use here, but I am hoarding it for the future. Malek Moazzam-Doulat has been willing to think about both Nietzsche and Wisdom, even together, along with whatever other peculiar notions I might run past him. And I owe Michael Kagan for his clear insights into Maimonides, and bewilderment, and dust motes floating in the light, whether or not they are secret.

I did not get to talk to Peter Manchester about this book. There were uncountably many times that I thought, "Oh, I'll have to ask Peter about . . . ," only to remember that this option is gone. I hope that I have done justice to his thoughts on the One and the Dyad, and that I have managed to pass along some of the joy he always shared in thinking about the cosmos.

My thinking about materialism generally, and Valentinianism in particular, owes much to two events organized by Jonathan Cahana-Blum during his postdoctoral research at the University of Aarhus on his project "Wrestling with Archons: Gnosticism as a Critical Theory of Culture." The first was a weeklong seminar of faculty and graduate students, the second an interdisciplinary conference juxtaposing ancient and contemporary thought; both were delightful. The conference led eventually to the volume *We and They: Decolonizing Greco-Roman and Biblical Antiquities*, which Jonathan and I (mostly Jonathan) edited, and which includes an early version of the chapter on limit.

The manuscript readers for this book, William Robert and Jeffrey Robbins, offered the perfect combination of helpfulness, precision, and enthusiasm—and did so with a promptness that is rare in our profession. My editor at Fordham, Tom Lay, somehow manages to be both easygoing and efficient, and it is a pleasure to work with him again. And finally, I am grateful to Alan Griffin, even though he always notices when I'm not really writing.

Notes

Introduction: New Materialism, Old Wisdom

1. René Descartes refers to matter as *res extensa*, extended things, in contrast to *res cogitans*, thinking things, to name the two kinds of substances in his metaphysics. See Descartes, *Meditations on First Philosophy*, trans. Donald Cress (Indianapolis: Hackett, 1993). "Brute thereness" is associated with the work of Charles Sanders Peirce, particularly with his difficult concept of "secondness," but I cannot be certain that Peirce is the first source of the phrase. See Peirce, *Collected Papers*, vol. 1, *Principles of Philosophy*, and vol. 2, *Elements of Logic* (Cambridge, Mass.: Harvard University Press, 1932). Martin Heidegger opens his *Introduction to Metaphysics* by declaring that the question "Why are there beings at all rather than nothing?" is "obviously the first of all questions"; Heidegger, *Introduction to Metaphysics*, trans. Gregory Fried and Richard Polt (New Haven, Conn.: Yale University Press, 2000), 1. Physicist Paul Dirac is generally credited with creating the contemporary concept of antimatter in a 1928 paper, "The Quantum Theory of the Electron," *Proceedings of the Royal Society of London, Series A, Containing Papers of a Mathematical and Physical Character* 117, no. 778 (1928): 610–24.

2. Tamsin Jones, "Introduction: New Materialism and the Study of Religion," in *Religious Experience and New Materialism: Movement Matters*, ed. Joerg Rieger and Edward Waggoner (London: Palgrave Macmillan, 2016), 1. For particularly good exemplars, see Catherine Keller and Mary-Jane Rubenstein, eds., *Entangled Worlds: Religion, Science, and New Materialisms* (New York: Fordham University Press, 2017).

3. Catherine Keller, *Intercarnations: Exercises in Theological Possibility* (New York: Fordham University Press, 2017), 61.

4. George Berkeley, *Three Dialogues Between Hylas and Philonous* (1713), ed. Robert M. Adams (Indianapolis: Hackett, 1979).

5. Astonishment, I increasingly think, has ethical import, and astonishment at materiality may be especially urgent. Though this connection shows up fairly often for me, my most focused consideration of it is in Karmen MacKendrick, *Divine Enticement: Theological Seductions*, in the chapter "Because Being Here Is So Much: Ethics as the Artifice of Attention" (New York: Fordham University Press, 2012), 101–40.

6. Again, Jones summarizes clearly: "Within this broader material turn, one of the defining elements of new materialism is the further rejection of an anthropocentrism central to much modern religious and theological thought. It is not much of a stretch to say that new materialism and religion, at first glance, seem to be a rather uncomfortable pairing"; Jones, "Introduction," 1. Jones cites Sonia Hazard, "The Material Turn in the Study of Religion," *Religion and Society: Advances in Research* 4 (2013): 58–78.

7. Levi Bryant, *The Democracy of Objects* (Ann Arbor, Mich.: Open Humanities, 2011), 20.

8. Diana Coole, "The Inertia of Matter and the Generativity of Flesh," in *New Materialisms: Ontology, Agency, and Politics*, ed. Diana Coole and Samantha Frost (Durham, N.C.: Duke University Press, 2010), loc. 1335, citing Giorgio Agamben, *The Open: Man and Animal*, trans. Kevin Attell (Stanford, Calif.: Stanford University Press, 2003), 39.

9. Genesis 1:26. Unless otherwise noted, all biblical citations are from the New Revised Standard Version.

10. Genesis 2:4–15.

11. Graham Harman, "The Road to Objects," *Continent* 3, no. 1 (2011): 171.

12. Harman, "Road," 178.

13. Timothy Morton, "Here Comes Everything," *Qui Parle* 19, no. 2 (Spring/Summer 2011): 165. Morton says that "OOO [object-oriented ontology] is a form of realism, not materialism," but the materialism to which he refers is that of a Cartesian substrate. See "Here Comes Everything," 179.

14. Samantha Frost and Diana Coole, "Introduction," in Frost and Coole, *New Materialisms*, location 355.

15. Genesis 2:19–20; Surah al Baqara, sec. 4, 2.31–33.

16. Patricia Cox Miller, "Adam, Eve, and the Elephants: Asceticism and Animality in Late Ancient Christianity," in *Ascetic Culture: Essays in Honor of Philip Rousseau*, ed. Blake Leyerle and Robin Young (Notre Dame, Ind.: Notre Dame University Press, 2013), and "The Pensivity of Animals II: Anthropomorphism," in Miller, *In the Eye of the Animal: Zoological Imagination in Ancient Christianity* (Philadelphia: University of Pennsylvania Press, 2018), 79–118.

17. Genesis 2:16–17.

18. See Plato, *Apology*, trans. Harold N. Fowler (London: William Heinemann, 1966), 21a–d. The skeptical period in the Academy, focusing on dialogue and the indeterminability of questions, runs from about 315–110 BCE, and was carried on by the Roman philosopher Cicero shortly afterward. Gregory Shaw's work is

particularly good on neo-Platonic theurgy; see, e.g., *Theurgy and the Soul: The Neoplatonism of Iamblichus* (Philadelphia: University of Pennsylvania Press, 2003).

19. Jones, "Introduction," 4.

20. Jane Bennett, "From Nature to Matter," in *Second Nature: Rethinking the Natural through Politics*, ed. Crina Archer, Laura Ephraim, and Lida Maxwell (New York: Fordham University Press, 2013), 149.

21. Bennett, "A Vitalist Stopover on the Way to New Materialism," in Frost and Coole, *New Materialisms: Ontology, Agency, and Politics*, locations 792–95.

22. Frost and Coole, "Introduction," location 101.

23. Frost and Coole, "Introduction," location 143.

24. Karen Barad, "Nature's Queer Performativity," *Qui Parle* 19, no. 2 (Spring/Summer 2011): 124.

25. Bruno Latour, *Pandora's Hope: Essays on the Reality of Science Studies* (Cambridge, Mass.: Harvard University Press, 1999), cited in Harman, "Road," 178.

26. Steven Shaviro, *Posthumanities: Universe of Things: On Speculative Realism* (Minneapolis: University of Minnesota Press, 2014), 77.

27. Morton, "Here Comes Everything," 165, with reference as well to Harman, "On Vicarious Causation," in *Collapse 2: Speculative Realism*, ed. Robin Mackay (Falmouth, UK: Urbanomic, 2007): 171–205.

28. For concise summaries, see Latour, "On Actor-Network Theory: A Few Clarifications," *Soziale Welt* 47 (1996): 369–81, and "On Recalling ANT," *Sociological Review* 47, no. 1 (1999): 15–25.

29. Genesis 1:26–28.

30. Genesis 2:15.

31. Genesis 3:1–7.

32. Karel van der Toorn, "Why Wisdom Became a Secret: On Wisdom as a Written Genre," in *Wisdom Literature in Mesopotamia and Israel*, ed. Richard J. Clifford (Atlanta: Society for Biblical Literature, 2007), 24–29.

33. Matthew Goff lists the Wisdom texts in the noncanonical literature from Qumran as 4QInstruction, *The Book of Mysteries*, 4Q184, 4Q185, and 4Q525; Goff, "Searching for Wisdom in and beyond 4QInstruction," in *Tracing Sapiential Traditions in Ancient Judaism: Traditions in Ancient Judaism*, ed. Hindy Najman, Jean-Sebastien Rey, and Eibert Tigchelaar (Leiden: Brill, 2016).

34. Ecclesiastes, written in the persona of a king, echoes the books of advice for princes that were popular in several ancient cultures. Jennifer L. Andruska offers extensive arguments that the Song of Songs, though difficult to categorize, shares some of the features of other advice-giving wisdom literature; Andruska, *Wise and Foolish Love in the Song of Songs* (Leiden: Brill, 2018).

35. Dianne Bergant, "Creation Theology in the Book of Job," in *America: The Jesuit Review* (January 10, 2008): unpaginated.

36. Proverbs 8:22–31.

37. Ronald Cox points out four New Testament texts in which evidence for the identification of Christ with the earlier figure of Wisdom is particularly strong. Three

are in Paul's letters—Hebrews 1, Colossians 1, and 1 Corinthians 1; the fourth is the prologue to the Gospel of John; Cox, *By the Same Word: Creation and Salvation in Hellenistic Judaism and Early Christianity* (Berlin: Walter de Gruyter, 2007), 2.

38. John 1:1–3.

39. John 1:14.

40. Virginia Burrus, *Ancient Christian Ecopoetics: Cosmologies, Saints, Things* (Philadelphia: University of Pennsylvania Press, 2018), 6.

41. Catherine Michael Chin, "Cosmos," in *Late Ancient Knowing: Explorations in Intellectual History*, ed. Catherine Michael Chin and Moulie Vidas (Berkeley: University of California Press, 2015), 100. Chin cites Heinz R. Schutte, *Weltseele: Geschichte und Hermeneutik* (Frankfurt: J. Knecht, 1993); and Alan Scott, *Origen and the Life of the Stars* (Oxford: Oxford University Press, 1991), chapter 8, "Origen and the Stars," 117–33.

42. An influential review of this material can be found in W. G. Lambert, *Babylonian Wisdom Literature* (Philadelphia: Eisenbrauns, 1960). See also Richard J. Clifford, "Introduction," in *Wisdom Literature in Mesopotamia and Israel* (Atlanta: Society for Biblical Literature, 2007), xi–xiv.

43. Craig Bartholomew and Ryan O'Dowd describe Wisdom theology as working together with a "natural wonder" at creation—though they do, I think erroneously, oppose this wonder to Platonic thought; Bartholomew and O'Dowd, *Old Testament Wisdom Literature: A Theological Introduction* (Downer's Grove, Ill.: InterVarsity Press Academic, 2018), 21.

44. See, for example, Elisa Uusimäki, who notes parallels between "Jewish wisdom pedagogy" and Hellenic *paideia*, as modes of education meant less to impart information than to form the person who learns; Uusimäki, "Spiritual Formation in Hellenistic Jewish Wisdom Teaching," in Rey, Najman, and Tigchelaar, *Tracing Sapiential Wisdom*, 58. Ishay Rosen-Zvi argues that the rabbinic innovation regarding Wisdom is not its identification with Torah, but "is the importation of wisdom into the *Beit Midrash*, and the equating of wisdom with the specific practices of the rabbinic house of study. Not just torah but halakh itself is now 'wisdom'"; Rosen-Zvi, "The Wisdom Tradition in Rabbinic Literature and Mishnah *Avot*," in Rey, Najman, and Tigchelaar, *Tracing Sapiential Wisdom*, 174–75.

45. See Rosen-Zvi, "Wisdom Tradition in Rabbinic Literature," 180.

46. Rosen-Zvi, "Wisdom Tradition in Rabbinic Literature," 172: "The tradition of identifying wisdom with torah may have found its first explicit attestation in Ben Sira (Sir 24:23, 39:1), but its roots are already in Deut 4:5–8. The rabbis inherited this tradition."

47. Maurice Gilbert, S. J., notes that one rabbinic text in particular, the third-century *Pirkei Avot*, is widely accepted as a work of Wisdom literature, and is written in the aphoristic style that characterizes wisdom as advice; Gilbert, "*Pirqé Avot* and Wisdom Tradition," in Rey, Najman, and Tigchelaar, *Tracing Sapiential Traditions*, 155. Gilbert points out parallels to the Greek *Sentences of Sextus*, the *Teachings of Silvanus* from Nag Hammadi, and the Syriac *Wisdom of the Egyptian Menander*,

whose non-Abrahamic writer nonetheless appears to be open to Judaism. All of these are roughly contemporaneous with *Pirkei Avot*.

48. Rosen-Zvi, "Wisdom Tradition in Rabbinic Literature," 190. Rosen-Zvi argues, "When we examine how Tannaitic literature uses the term 'wisdom,' we understand that the reason is not the lack of identification of Torah and wisdom but, quite to the contrary, its omnipresence. . . . Torah does not have to be wisdomized, it is, by definition, wisdom" (174).

49. Ismo Dunderberg, "Valentinus and His School," *Revista Catalana de Teologia* 37, no. 1 (2012): 135. Dunderberg recommends the account of this group in David Brakke, *The Gnostics: Myth, Ritual, and Diversity in Early Christianity* (Cambridge, Mass.: Harvard University Press), 2010. The group gets its name because members regarded themselves as spiritual descendants of Seth, the third son of Adam. Seth does not make extensive appearances in canonical scripture. His birth is mentioned in Genesis 4:25, and that of his son Enosh at 4:26; the same series is reiterated at 5:3–5:8, with mention of Adam and Seth both living to great ages. Beyond this, he appears only in genealogical lists at 1 Chronicles 1:1 and Luke 3:38, and is mentioned in Sirach 49:16 as being honored, but not as highly as Adam.

50. See John D. Turner, *Sethian Gnosticism and the Platonic Tradition* (Peeters, Belgium: Le Presses de l'Université Laval, 2006), 257. Turner, who has worked extensively on Middle Platonism and mathematical theology and philosophy, offers a clear summary in "Sethianism," in *The Encyclopedia of Ancient History*, ed. Roger S. Bagnall, Kai Brodersen, Craige B. Champion, Andrew Erskine, and Sabine R. Huebner (London: Blackwell, 2013), 6176–79.

51. Ismo Dunderberg, "Recognizing the Valentinians—Now and Then," in *The Other Side: Apocryphal Perspectives on Ancient Christian "Orthodoxies*," ed. Tobias Niklas, Candida R. Moss, Christopher Tuckett, and Joseph Verhayden (Göttingen: Vanderhoeck and Ruprecht, 2017), 40.

52. Dunderberg, "Recognizing the Valentinians," 46–52.

53. Michael A. Williams offers a carefully detailed list of scholarly works identifying Gnosticism with the view that the created world is evil, in "A Life Full of Meaning and Purpose: Demiurgical Myths and Social Implications," in *Beyond the Gnostic Gospels: Studies Building on the Work of Elaine Pagels*, ed. Eduard Iricinschi, Lance Jenott, Nicola Denzey Lewis, and Philippa Townsend (Tübingen: Mohr Siebeck, 2013), 21–22n6. His citations include Majella Franzmann, "A Complete History of Early Christianity: Taking the 'Heretics' Seriously," *Journal of Religious History* 29 (2005); Bart D. Ehrman, "Christianity Turned on Its Head: The Alternative Vision of the Gospel of Judas," in *The Gospel of Judas from Codex Tchacos*, ed. Rodolphe Kasser et al. (Washington, D.C.: National Geographic, 2006); Edmondo Lupieri, *The Mandaeans: The Last Gnostics*, trans. Charles Hindley (Cambridge: Eerdmans, 2002); Giovanni Filoramo, *A History of Gnosticism*, trans. Anthony Alcock (Oxford: Basil Blackwell, 1990); Sebastian Moll, *The Arch-Heretic Marcion* (Tübingen: Mohr Siebeck, 2010). Among the most foundational of these identifications is Hans Jonas, *The Gnostic Religion: The Message of the Alien God and the Beginnings of Christianity*

(Boston: Beacon Press, 1963). We may add to these the relatively recent example of Roger E. Olsen, *The Story of Christian Theology: Twenty Centuries of Tradition and Reform* (Downer's Grove, Ill.: InterVarsity Press, 1999), esp. 36f.

54. Wisdom 7:24.

1. Complex Truth: Myths, Facts, and Matter

1. Alfred North Whitehead, *Process and Reality* (New York: Free Press, 1978), 39.

2. Diogenes Laërtius, *Lives and Opinions of Eminent Philosophers, Including the Biographies of the Cynics and the Life of Epicurus*, trans. Robert Drew Hicks (New York: Gottfried and Fitz, 2014), 2.4.

3. Diogenes Laërtius, *Lives and Opinions*, 1.14.

4. Parmenides, *On Nature (Peri Physeos)*, "The Way of Truth," trans. Allan Randall, 1996, accessed March 27, 2021, at http://www.allanrandall.ca/Parmenides.html, B 8.34–37.

5. "Seems to have" is important. Because we are missing some of Parmenides's major text *On Nature*, and because the context of his practice of teaching is lost to us, we cannot quite be sure what he is up to. Myth makers, particularly, are often up to subtler things than they seem at first to be. My remarks on Parmenides are grounded in what is by now a fairly mainstream and received tradition, but other readings are certainly possible and possibly more interesting.

6. Gregory Shaw, "Theurgy and the Platonist's Luminous Body," in *Practicing Gnosis: Ritual, Magic, Theurgy, and Liturgy in Nag Hammadi, Manichaean and Other Ancient Literature; Essays in Honor of Birger A. Pearson*, ed. April DeConick, Gregory Shaw, and John D. Turner (Leiden: Brill, 2013), 537: "Our Plato is a fiction, a caricature, a venerable man of straw. The great philosopher now disparaged by post-modern critics as the architect of metaphysical dualism is not the Plato recognized by the philosophers of late antiquity. . . . This Plato—familiar to us—would scarcely have been recognizable to the Platonists of antiquity. He would have been scarcely recognizable because the Plato we have inherited is an invention of our own habits of thought, and the dualism we attribute to him reflects our own existential estrangement from the divinity of the world."

7. Diogenes Laërtius, *Lives and Opinions*, 1.22.

8. Diogenes Laërtius notes several other astronomical contributions attributed to him: the discoveries of the constellation Ursa Minor, of the period between solstices, and of the determination of the sizes of the sun and moon; *Lives and Opinions*, 1.24–25.

9. Diogenes Laërtius, *Lives and Opinions*, 1.26.

10. Diogenes Laërtius, *Lives and Opinions*, 1.26.

11. Diogenes Laërtius, *Lives and Opinions*, 1.26.

12. Aristotle, *Metaphysics*, trans. Hugh Tredenick (London: William Heinemann, 1989), 1.983b.

13. Aristotle, *Metaphysics* 983b2.

14. Aristotle, *Metaphysics* 981b1.
15. Aristotle, *Metaphysics* 981b2.
16. Aristotle, *Metaphysics* 983a1; cf. Aristotle, *Physics*, trans. Robin Waterfield (Oxford: Oxford University Press, 2008), 2.3, 2.7.
17. Anaximander proposed *apeiron*, the unlimited or unformed, as the first substance, and Anaximenes gave primacy to *aer*, air in the sense of mist or vapor.
18. The *Theogony* opens, "From the Heliconian Muses let us begin to sing, who hold the great and holy mount of Helicon, and dance on soft feet about the deep-blue spring and the altar of the almighty son of Cronos, and, when they have washed their tender bodies in Permessus or in the Horse's Spring or Olmeius, make their fair, lovely dances upon highest Helicon and move with vigorous feet"; Hesiod, *Theogony*, trans. Hugh G. Evelyn-White (London: William Heinemann, 1914). Homer's *Odyssey* likewise invokes the muses at its opening: "Tell me, O Muse, of the man of many devices, who wandered full many ways after he had sacked the sacred citadel of Troy"; Homer, *The Odyssey*, trans. A. T. Murray (London: William Heinemann, 1919).
19. Homer, *The Iliad*, trans. Robert Fagles (New York: Penguin, 1998), 14.245. Patricia Curd discusses Hesiod's cosmogony and its difference from philosophical accounts in her introduction to *A Presocratics Reader*, ed. Patricia Curd (Indianapolis: Hackett, 2011).
20. Plato, *Theaetetus*, trans. John McDowell (Oxford: Oxford University Press, 2014), 152d–e, citing Homer, *The Iliad* 2.14.201, 302.
21. Aristotle, *Metaphysics* 983a, 982b.
22. "There are some who think that the men of very ancient times, long before the present era, who first speculated about the gods, also held this same opinion about the primary entity. For they represented Oceanus and Tethys to be the parents of creation"; Aristotle, *Metaphysics* 983b. Aristotle also notes that both Hesiod and Parmenides "assume Love or Desire as a first principle in things"; Aristotle, *Metaphysics* 984b. Aristotle refers to Hesiod, *Theogony* 116–20 and to Parmenides, Frag. 13.
23. Aristotle, *Metaphysics* 1.983a–b.
24. M. R. Wright, "Presocratic Cosmogonies," in *The Oxford Handbook of Presocratic Philosophy*, ed. Patricia Curd and Daniel W. Graham (Oxford: Oxford University Press, 2008), 414. Guthrie writes, "The explanation of nature is to be sought within nature itself"; W. K. C. Guthrie, *A History of Greek Philosophy*, vol. 1, *The Earlier Presocratics and the Pythagoreans* (Cambridge: Cambridge University Press, 1962), 45.
25. See Jaap Mansfeld, "Aristotle and Others on Thales, or The Beginnings of Natural Philosophy," *Mnemosyne* 38, no. 1–2 (1985): 111.
26. Aristotle, *Metaphysics* 1000a.
27. Aristotle, *Metaphysics* 1000a.
28. Mansfeld, "Aristotle and Others," 122.
29. Daniel Cossins, "Why We Like to Know Useless Stuff," *New Scientist* 29, March 2017, accessed March 27, 2021, at https://www.newscientist.com/article/mg23431190-700-knowledge-why-we-like-to-know-useless-stuff/.

30. Curd cites Xenophon and Alcmaeon in discussing this epistemological shift: "Xenophanes of Colophon specifically rejects this justification. 'By no means,' he says (21B18), 'did the gods intimate all things to mortals from the beginning, but in time, by inquiring, they discover better.' In rejecting divine authority for their claims, the Presocratics invite inquiry into the sources of human knowledge. A tantalizing mention of this problem appears in a fragment from Alcmaeon, who echoes Homer's claims that the gods know all things, but apparently offers a more pessimistic outlook for humans: 'Concerning the unseen, the gods have clarity, but it is for men to conjecture from signs' (DK24B1)"; Curd, *Presocratics Reader*, 5.

31. Curd, "Presocratic Philosophy," *Stanford Encyclopedia of Philosophy*, 2016, accessed March 27, 2021, at https://plato.stanford.edu/entries/presocratics/.

32. Tim Whitmarsh, *Battling the Gods: Atheism in the Ancient World* (New York: Vintage, 2015), 54.

33. Richard Rojcewicz, "Everything Is Water," *Research in Phenomenology* 44 (2014): 195.

34. Frederic May Holland, *The Rise of Intellectual Liberty from Thales to Copernicus* (New York: H. Holt, 1885), 4.

35. Stephen White, "Milesian Measures: Time, Space, and Matter," in Curd and Graham, *Oxford Handbook of Presocratic Philosophy*, 90.

36. This is less odd than it might appear, since the Society of Jesus has something of a history of irksome questioning. The order was suppressed in many countries, and finally by papal edict, in the late eighteenth century, though probably for political rather than theological reasons. It was fully restored in 1815.

37. See René Descartes, Letter to his translator, 1647, in Descartes, *The Philosophical Writings of Descartes*, vol. 1, *Principles of Philosophy*, trans. J. Cottingham, R. Stoothoff, and D. Murdoch (Cambridge: Cambridge University Press, 1985), 179–89.

38. Descartes, Letter of Dedication, in *Meditations on First Philosophy*, trans. Donald Cress (Indianapolis: Hackett, 1993), 1–4. To be sure, this letter of dedication is also (or perhaps chiefly) meant to protect Descartes's *Meditations* in a risky political climate.

39. The arguments for the existence of God are all well-known medieval arguments and not new proofs according to reason. The proof for the soul's immortality is that bodies disintegrate after death. Souls do not have parts, so they cannot fall apart, so they must go on forever. Plato makes a similar point in the *Phaedo*, in *Plato: Five Dialogues; Euthyphro, Apology, Crito, Meno, Phaedo*, trans. G. M. A. Grube, rev. John M. Cooper (Indianapolis: Hackett, 2002), 83a–b; Descartes, *Meditations*, Book 3 (proofs of God's existence), Book 6 (proof of the soul's immortality).

40. William Bristow, "Enlightenment," *Stanford Encyclopedia of Philosophy*, 2017, accessed December 7, 2019, at https://plato.stanford.edu/entries/enlightenment/.

41. Bristow, "Enlightenment."

42. Immanuel Kant, "Answer to the Question: What Is Enlightenment?," trans. Ted Humphrey, in *Perpetual Peace and Other Essays* (Indianapolis: Hackett, 1983), §§1–2.

43. Kant, "What Is Enlightenment?" §§ 3–4.

44. Theodor Adorno and Max Horkheimer, *Dialectic of Enlightenment: Philosophical Fragments*, trans. Edmund Jephcott (Stanford, Calif.: Stanford University Press), 1.

45. Adorno and Horkheimer, *Dialectic*, 11.

46. Epicurus, "Principal Doctrines," in *The Essential Epicurus: Letters, Principal Doctrines, Vatican Sayings, and Fragments*, trans. Eugene O'Connor (Buffalo, N.Y.: Prometheus, 1993).

47. Adorno and Horkheimer, *Dialectic*, 2.

48. Aristotle, *Metaphysics* 980a.

49. Adorno and Horkheimer, *Dialectic*, 2.

50. Adorno and Horkheimer, *Dialectic*, 2.

51. Adorno and Horkheimer, *Dialectic*, 3.

52. Adorno and Horkheimer, *Dialectic*, 3. Cf. Jane Bennett's descriptions of enchantment in *Vibrant Matter: A Political Ecology of Things* (Durham, N.C.: Duke University Press, 2010), and *The Enchantment of Modern Life: Attachments, Crossings, and Ethics* (Princeton, N.J.: Princeton University Press, 2001).

53. Adorno and Horkheimer, *Dialectic*, 6.

54. Marjorie Garber, *The Muses on Their Lunch Hour* (New York: Fordham University Press, 2017), 30: "Few scientists seem to have premised their experiments on the foundation of humanistic work (despite the existence of some learned books on art by scientists)." In a now-famous Twitter exchange in 2017, Neil deGrasse Tyson, an excellent science popularizer, wrote, "In school, rarely do we learn how data become facts, how facts become knowledge, and how knowledge becomes wisdom." The response from the humanities scholars who study exactly these questions was speedy and plentiful, but deGrasse Tyson's claim shows something of the ease with which nonscience disciplines are disregarded. See https://twitter.com/neiltyson/status/904861739329708034.

55. Particularly engaging among work that crosses theological depth with contemporary physics is that of Catherine Keller: see especially *The Cloud of the Impossible: Negative Theology and Planetary Entanglement* (New York: Columbia University Press, 2015) and Mary-Jane Rubenstein: see especially *Worlds without End: The Many Lives of the Multiverse* (New York: Columbia University Press, 2014) and *Pantheologies: Gods, Worlds, Monsters* (New York: Columbia University Press, 2018).

56. Bruno Latour, *On the Modern Cult of the Factish Gods* (Durham, N.C.: Duke University Press, 2010).

57. The sense of "myth" as meaning "untrue story, rumor" is from 1840; *Online Etymology Dictionary*, accessed December 7, 2019, at https://www.etymonline.com/word/myth.

58. This infamous claim came from Sen. James Inhofe of Oklahoma in 2015.

59. For the original use of "alternative facts," see interview with Kellyanne Conway, *Meet the Press*, January 22, 2017. Wikipedia offers a good summary at https://en.wikipedia.org/wiki/Alternative_facts.

60. The problem is also Fox News and the rest of the Murdoch empire, but this is beyond the scope of the present work.

61. Mansfeld, "Aristotle and Others," 122, 127.

62. Mansfeld, "Aristotle and Others," 125.

63. Aristotle, *On the Heavens*, trans. W. K. C. Guthrie (Cambridge, Mass.: Harvard University Press, 1939), 249a28.

64. Aristotle, *De Anima (On the Soul)*, trans. R. D. Hicks (Cambridge: Cambridge University Press, 1907), 405a19; see also Diogenes Laërtius, *Lives and Opinions* 1.24.

65. This is Diogenes's phrasing, in *Lives and Opinions*, 1.27. Aristotle writes that for Thales all things are "full of gods"; *De Anima* 411a7.

66. Hippolytus, *The Refutation of All Heresies* 1.1, trans. J. H. MacMahon, in *Nicene and Ante-Nicene Fathers*, vol. 5, ed. A. Cleveland Coxe, Alexander Roberts, and James Donaldson (Buffalo, N.Y.: Christian Literature, 1886).

67. Patricia F. O'Grady, *Thales of Miletus: The Beginnings of Western Science and Philosophy* (New York: Routledge, 2016), 110–11.

68. Aristotle, *Physics* 3.4 203b10–15.

69. Hippolytus, *Refutation* 1.6.

70. Aristotle, *Metaphysics* 1072a.

71. Aristotle, *Metaphysics* 1074b; see also *Physics* 8.8–10.

72. O'Grady, *Thales*, 111.

73. Whitmarsh, *Battling the Gods*, 59.

74. Curd, "Presocratic Philosophy."

75. Diogenes Laërtius, *Lives and Opinions* 1.27: "His doctrine was that water is the universal primary substance, and that the world is animate and full of divinities."

76. Whitmarsh, *Battling the Gods*, 63. Hippo is pre-Socratic; his work probably dates to the fifth century BCE.

77. Aristotle, *Metaphysics* 984a. Cf. *De Anima* 405a, where Aristotle calls Hippo "superficial."

78. Diogenes Laërtius, *Lives and Opinions* 1.4.52.

79. Diogenes Laërtius, *Lives and Opinions* 1.4.54. Theodorus dates to the fourth century BCE. Poor Bion himself is described as "a shifty character, a subtle sophist, and one who had given the enemies of philosophy many an occasion to blaspheme, while in certain respects he was even pompous and able to indulge in arrogance"; 1.4.48.

80. Whitmarsh, *Battling the Gods*, 124. Athenagoras writes, "With reason did the Athenians adjudge Diagoras guilty of atheism, in that he not only divulged the Orphic doctrine, and published the mysteries of Eleusis and of the Cabiri, and chopped up the wooden statue of Hercules to boil his turnips, but openly declared that there was no God at all"; Athenagoras, *A Plea for the Christians*, trans. B. P. Pratten, in *Ante-Nicene Fathers*, vol. 2, ed. A. Cleveland Coxe, Alexander Roberts, and James Donaldson (Buffalo, N.Y.: Christian Literature, 1885), chap. 4.

81. Mary-Jane Rubenstein offers an excellent, detailed history of the anxiety that anything like pantheism creates for philosophical and religious thought, in *Pantheologies*.

82. Rubenstein, *Pantheologies*, 12.

83. Xenophanes, fragments B12, B16, and B15, in Curd, *Presocratics Reader*.
84. Denys Turner, "How to Be an Atheist," *New Blackfriars* 83, no. 977/978 (July–August 2002): 320.
85. Denys Turner, "How to Be an Atheist," 321.
86. Plato, *Symposium*, trans. Alexander Nehamas and Paul Woodruff (Indianapolis: Hackett, 1989), 211a–b.
87. "Scholarly consensus establishes Plato as the beginning of systematic reflection on myth"; Dexter E. Callender Jr., "Myth and Scripture: Dissonance and Convergence," in *Myth and Scripture: Contemporary Perspectives on Religion, Language, and Imagination*, ed. Dexter E. Callender Jr. (Atlanta: Society for Biblical Literature, 2014), 40–41, with reference to Andrew Von Hendy, *The Modern Construction of Myth* (Bloomington: Indiana University Press, 2002).
88. Robert Fowler, "Mythos and Logos," *Journal of Hellenic Studies* 131 (2011): 50. Fowler cites the *Gorgias*, *Timaeus*, *Republic*, *Cratylus*, and *Phaedo*.
89. Plato, *Sophist*, trans. Harold Fowler (London: William Heinemann, 1921), 242c; see also 242d–e.
90. Plato, *Republic*, esp. 595a–600c.
91. Plato, *Republic* 601b–7c.
92. Plato, *Republic* 607c.
93. Fowler, "Mythos and Logos," 50. Fowler notes that Aristotle's argument is similar, citing *Metaphysics* 699a27–32.
94. Callender, *Myth and Scripture*, 28.
95. Plato, *Republic* 377a.
96. Fowler, "Mythos and Logos," 58.
97. Fowler provides a good discussion of the Sophists' use of *logos* and Plato's alteration of it; see especially "Mythos and Logos," 57–58.
98. Spyros Orfanos, "Mythos and Logos," *Psychoanalytic Dialogues* 16 no. 4 (July–August 2006): 484.
99. Catalin Partenie notes that Plato makes use of traditional myths (sometimes modified or combined for his own purposes), myths that he invents (though they may make use of traditional characters), and philosophical doctrines that he himself nonetheless calls myths. He provides meticulous evidence for each; see Partenie, "Introduction," in *Plato's Myths*, ed. Partenie (Cambridge: Cambridge University Press, 2009), esp. 1–4.
100. Plato, *Republic* 369b–72a.
101. Plato, *Republic* 372b.
102. Plato, *Republic* 614b–21b.
103. Catalin Partenie, "Plato's Myths," *Stanford Encyclopedia of Philosophy*, 2018. Accessed March 27, 2021, at https://plato.stanford.edu/entries/plato-myths/.
104. Shaw, "Theurgy and the Platonist's Luminous Body," 538. See Iamblichus, *The Life of Pythagoras*, in *The Pythagorean Source Book and Library*, trans. Kenneth Sylvan Guthrie (Alpine, N.J.: Platonist Press, 1919), chap. 15.
105. Shaw, "Theurgy and the Platonist's Luminous Body," 538.

106. Shaw, "Theurgy and the Platonist's Luminous Body," 539, especially of Iamblichus. Interestingly, Diogenes actually credits Pythagoras as one of two originators of "philosophy, the pursuit of wisdom," the other being Thales's pupil Anaximander, in *Lives and Opinions* 1.13. Diogenes says of Plato, "In his doctrine of sensible things he agrees with Heraclitus, in his doctrine of the intelligible with Pythagoras, and in political philosophy with Socrates"; Diogenes Laërtius, *Lives and Opinions* 3.8.

107. Plato, *Symposium* 198c–99a.

108. At *Symposium* 205e, Diotima makes a reference to the story that Aristophanes has just told, a fact that Aristophanes tries to point out at 212c.

109. Plato, *Symposium* 209e–10a.

110. Plato, *Symposium* 211b.

111. Shaw, "Theurgy and the Platonist's Luminous Body," 540, citing Jean Trouillard, *Le mystagogie de Proclos* (Paris: Les Belles Lettres, 1982), 233–34.

112. Friedrich Nietzsche, *The Gay Science*, trans. Walter Kaufmann (New York: Vintage, 1974), §84. Note also William Franke, "Poetry, Prophecy, and Theological Revelation," *Oxford Research Encyclopedia in Religion* (Online) (Oxford: Oxford University Press, 2016), 4.

113. Nietzsche, *Gay Science*, §84.

114. Lynne Huffer, *Mad for Foucault: Rethinking the Foundations of Queer Theory* (New York: Columbia University Press, 2009), 259–60.

115. William Franke notes that poetry, including but not limited to mythic poetry, "makes sense in another way than that to which we are accustomed," making "it possible to think and feel in correspondence with indeterminacies and infinites created and opened [by] up the negative employment of words"; Franke, "Poetry, Prophecy, and Theological Revelation," 6.

116. Callender, "Myth and Scripture," 44.

117. Athanasius, "Letter to Marcellinus," in *The Life of Antony and The Letter to Marcellinus*, trans. Robert C. Gregg (Mahwah, N.J.: Paulist Press, 1980), 101–30.

118. Partenie, *Plato's Myths*, 19, citing C. J. Rowe, "Myth, History, and Dialectic in Plato's *Republic* and *Timaeus-Critias*," in *From Myth to Reason? Studies in the Development of Greek Thought*, ed. R. Buxton (Oxford: Oxford University Press, 1999), 255.

119. Michel Foucault, *The Hermeneutics of the Subject: Lectures at the Collège de France 1981–1982*, trans. Graham Burchell (New York: Palgrave Macmillan, 2005), 17.

120. Adorno and Horkheimer, *Dialectic*, 2.

121. Jean-Luc Nancy, *The Pleasure in Drawing*, trans. Philip Armstrong (New York: Fordham University Press, 2013), 41.

122. Maurice Blanchot, *The Infinite Conversation*, trans. Susan Hanson (Minneapolis: University of Minnesota Press, 1992), 314.

2. Adam's Skin: The Strangely Bounded Primal Person

1. Maurice Blanchot, *The Writing of the Disaster*, trans. Ann Smock (Lincoln: University of Nebraska Press, 2015), 117.

2. Genesis 1:1. Unless otherwise noted, all biblical translations are from the New Revised Standard Version.

3. Genesis 1:1–2.

4. Genesis 1:4–5, 14–18.

5. Genesis 1:6–8, 9–10.

6. Abraham Benish, *Jewish School and Family Bible*, vol. 1, *Pentateuch* (London: Longman, Brown, Green, and Longman's, 1852), 1. In the well-known King James and Douay-Rheims versions, God "divided" rather than "separated."

7. See Genesis 1:29–31.

8. Philo, *On the Creation of the World*, in *The Works of Philo: Complete and Unabridged*, trans. Charles D. Yonge (London: H. G. Bohn, 1854–90), 1.9.33.

9. Genesis 1:20–25.

10. Genesis 1:26–27.

11. Genesis 1:29–30.

12. Genesis 1:31.

13. As Elliot R. Wolfson writes of Kabbalistic readings of Genesis, "The origin, signified by alef, paradoxically, cannot be the beginning, for the beginning bears the mystery of beit, the dyad, the second that is first." This play between first and dyad will be still more significant in the next chapter; Wolfson, *Language, Eros, Being: Kabbalistic Hermeneutics and Poetic Imagination* (New York: Fordham University Press, 2004), 196.

14. Genesis 2:5–6.

15. The source of the first narrative is probably dated to the post-exilic period (specifically the sixth to fifth centuries BCE); the second is likely from the tenth to ninth centuries BCE—though there is as much debate as consensus on dating scripture.

16. Genesis 2:7.

17. Genesis 2:21–22. The animals that Adam named earlier were intended to be his companions, but none is a suitable partner, leading God to make the woman instead; Genesis 2:18–21.

18. One of the best known and most entertaining fusions of the two creation tales is the story of Lilith, which can be found in *The Alphabet of Ben Sirach*, 78. The *Alphabet* was inspired by Sirach (Ecclesiasticus), and it includes a series of moral-sounding aphorisms, but it is irreverent in tone and also includes folk tales, bawdy humor, and a lot of fable-like stories about animals. I do not know of a complete English translation, but a partial translation along with an account of the work is in A. E. Cowley and A. Neubauer, *The Original Hebrew of a Portion of Ecclesiasticus* (Oxford: Clarendon, 1897).

19. *Targum Pseudo-Jonathan*, as *The Targum of Jonathan ben Uzziel*, composed ca. 150–250, trans. J. W. Etheridge (London: Longman, Green, Longman and Roberts, 1862), on Genesis 1:27.

20. This is not quite the case in more orthodox Christianity, though Augustine attributes androgyny to the human mind and notes how worried some scholars are by the possibility of an androgynous body; Augustine, *On the Trinity* 12.7, in Philip Schaff, ed., *Nicene and Post-Nicene Fathers*, Series 1 (London: T & T Clark, 1887), 3:324.

21. *Vayikra (Leviticus) Rabbah* 14, composed ca. 500 CE, trans. Rabbi Mike Feuer, accessed March 27, 2021, at http://www.sefaria.org/Vayikra_Rabbah.14?lang=en. In all notes, where the URL differs between the note and bibliographic references, the note reference is to the cited point in the text and the bibliographic reference is to the starting point of the text.

22. *Genesis Rabbah* 8:1, composed ca. 500 CE, Sefaria Community Translation, accessed March 27, 2021, at https://www.sefaria.org/Bereishit_Rabbah.8?lang=bi; Plato, *Symposium*, trans. Alexander Nehamas and Paul Woodruff (Indianapolis: Hackett, 1989), 189c–93e.

23. *The Apocalypse of Adam* 1:2–3, translated by G. MacRae, in *The Old Testament Pseudepigrapha*, ed. James H. Charlesworth (Peabody, Mass.: Hendrickson, 1983), 1:712.

24. "If the female had not separated from the male, the female and the male would not have died. The separation of male and female was the beginning of death. Christ came to heal the separation that was from the beginning and reunite the two, in order to give life to those who died through separation and reunite them"; *The Gospel of Philip*, trans. Marvin Meyer, in *The Nag Hammadi Scriptures*, ed. Marvin Meyer (New York: HarperCollins, 2008), 70, 9–10.

25. See the *Zohar*, trans. with commentary by Daniel C. Matt (Stanford, Calif.: Stanford University Press, 2006), 1:237a, n. 663.

26. Genesis 1:26.

27. Ibn Ezra on Genesis 1:26, composed ca. 1155–65 CE, Sefaria Community Translation, accessed December 9, 2019, at https://www.sefaria.org/Ibn_Ezra_on_Genesis.1.26?lang=bi.

28. Judah Halevi, *Sefer Kuzari* 4:3, composed ca. 1120–30 CE, trans. Hartwig Hirschfeld, accessed March 27, 2021, at https://www.sefaria.org/Sefer_Kuzari.4.3?lang=bi&with=all&lang2=en.

29. Rashi (Rabbi Shlomo Yitzchaki) on Genesis 1:26, in *Pentateuch with Rashi's Commentary*, composed ca. 1075–1105 CE, trans. M. Rosenbaum and A. M. Silbermann, accessed March 27, 2021, at https://www.sefaria.org/Rashi_on_Genesis?lang=bi.

30. *Midrash Tanchuma-Yelammedenu*, composed ca. 500–800 CE, trans. Samuel Berman (Hoboken, N.J.: KTAV, 1996), Pekudei, Siman 3, accessed March 27, 2021, at https://www.sefaria.org/Midrash_Tanchuma%2C_Pekudei.3?lang=bi.

31. Nachmanides (Moses ben Nachman, also known as Ramban) on Genesis, 1:26, composed ca. 1246–86 CE, Sefaria Community Translation, accessed December 8,

2019, at https://www.sefaria.org/Ramban_on_Genesis.1.26.1?lang=bi&with=all&lange2=en. Cf. Rashi on Genesis 2:7.

32. *Gospel of Philip* 71:16–20.

33. Jacob ben Asher, *Tur HaAroch*, on Genesis 1:26, composed early fourteenth century CE, trans. Eliyahu Munk (Jerusalem: Urim, 2005), accessed March 27, 2021, at https://www.sefaria.org/Tur_HaAroch%2C_Genesis.1.24?lang=bi.

34. Genesis 3:19: "By the sweat of your face you shall eat bread until you return to the ground, for out of it you were taken; you are dust, and to dust you shall return."

35. 2 Enoch 30:10, as *The Book of the Secrets of Enoch*, in *The Forgotten Books of Eden*, trans., ed. Rutherford H. Platt (New York: Alpha House, 1926).

36. *Genesis Rabbah* 6:3. The Talmudic tractate *Sanhedrin* offers more detail about the dusty sources of the first person: "His torso was fashioned from dust taken from Babylonia and his head was fashioned from dust taken from Eretz Yisrael, . . . and his limbs were fashioned from dust taken from the rest of the lands in the world. With regard to his buttocks, Rav Aḥa says, They were fashioned from dust taken from Akra De'agma, on the outskirts of Babylonia"; *Sanhedrin* 38b, composed ca. 450–550 CE, trans. Adin Even-Israel Steinsaltz, accessed March 27, 2021, at https://www.sefaria.org/Sanhedrin?lang=bi&p2=Sanhedrin.38b.2&lang2=en. The tractate goes on to explain that from Adam's creation to the exile from Eden, only half a day elapses.

37. *Genesis Rabbah* 14:1. In many of these stories, sometimes implicit, is a background of destruction: God has made many worlds, and because humans keep screwing up, has destroyed each one. To include the possibility of atonement, therefore, is also to make it possible for the world and the humans on it to endure.

38. Rashi on Genesis 2:7: "He gathered his dust (i.e., that from which he was made) from the entire earth—from its four corners—in order that wherever he might die, it should receive him for burial (*Midrash Tanchuma* 2:11:3). Another explanation: He took his dust from that spot on which the Holy Temple with the altar of atonement was in later times to be built of which it is said, (Exodus 20:24) 'An altar of earth thou shalt make for Me' saying, 'Would that this sacred earth may be an expiation for him so that he may be able to endure' (*Genesis Rabbah* 14:8)."

39. *Zohar* 3:31a.

40. *Chronicles of Jerahmeel* 6.3, composed fourteenth century, trans. Moses Gaster (London: Oriental Translation Fund, 1899).

41. *Jerahmeel* 6:3.6. The earth's claim is a reference to Genesis 3:17, in which God curses Adam after the latter has eaten the forbidden fruit.

42. *Jerahmeel* 6:3.7. The soul is also the breath; God breathes in life, and it is when breathing ceases that the soul is gone. "God kneaded and moulded the dust for the first man in a pure place, He covered him with skin and sinews, and gave to it a human shape, but there was not yet any breath or soul in it. What did God do? He breathed with the breath of His mouth, and thrust the soul into him, as it is said, 'And He breathed in his nostrils the breath of life.'" Cf. *Genesis Rabbah* 14:9; Genesis 1:30, 2:7.

43. Clement of Alexandria, *Excerpta Ex Theodoto of Clement of Alexandria*, composed ca. 175–200 CE, trans. Robert Pierce Casey (London: Christophers, 1934), Fragment 50.

44. *Excerpta ex Theodoto*, Fragment 51.

45. *Excerpta ex Theodoto*, Fragment 53.

46. Tertullian, *On the Resurrection*, ca. 160–220 CE, trans. Peter Holmes, in *Ante-Nicene Fathers*, ed. A. Cleveland Coxe, Alexander Roberts, and James Donaldson, vol. 3 (Buffalo, N.Y.: Christian Literature Publishing, 1885), chap. 6. Tertullian also insists that God has indeed comingled "the breath of His own Spirit" with the "vilest sheath" of the flesh; "and so intimate is the union, that it may be deemed to be uncertain whether the flesh bears about the soul, or the soul the flesh; or whether the flesh acts as *apparitor* to the soul, or the soul to the flesh"; *On the Resurrection*, chap. 7.

47. Tertullian, *Resurrection*, chap. 7.

48. Tertullian, *Resurrection*, chap. 7.

49. *Jerahmeel* 6:3.12; *Genesis Rabbah* 8. *Leviticus Rabbah* 14:7 offers a very similar passage:

> When God created the first man, from one end of the world to the other end, He created him to fill the entire world. How do we know from east to west? As it says, "You created me back to east *(kedem).*" How do we know from north to south? As it says, [Deuteronomy 4:32] "From the ends of the heavens to the ends of the heavens." How do we know that it was the expanse of the world? As it says, "And You laid your hand on me."

David Brakke notes the parallel of "the Jewish Shiur Qomah or measurement-of-the-body literature, which gives astoundingly large measurements of parts of God's body, along with esoteric names for them"; Brakke, "The Body as/at the Boundary of Gnosis," *Journal of Early Christian Studies* 17, no. 2 (2009): 208, citing Pieter W. van der Horst, "The Measurement of the Body: A Chapter in the History of Ancient Jewish Mysticism," in *Effigies Dei: Essays on the History of Religions*, ed. Dirk van der Plas, Studies in the History of Religions (Supplements to Numen) 51 (Leiden: Brill, 1987), 60.

50. Alexander Altmann, *Studies in Religious Philosophy and Mysticism* (Ithaca, N.Y.: Cornell University Press, 1969), 20–21. Altmann provides a number of useful references: for Pythagoras, *Vita Anonymi of Pythagoras*, in *Photi Bibliotheca*, cod. 249; for Democritus, fragment 4; for Aristotle, *Physics*, 252b, 26–27; for Plato, *Timaeus*, 30c–d; for Hermeticism, the *Emerald Tablet*; for Marcus Manilius, book 4 of his *Astronomica*; for Firmicus Maternus, section 3 of his *Mathesis*.

51. Ibn Ezra on Genesis 1:26.

52. *Midrash Tanchuma*, Pekudei, Siman 3:17, accessed March 27, 2021, at https://www.sefaria.org/Midrash_Tanchuma%2C_Pekudei.3?lang=bi, Pekudei 3:17. Raphael Patai notes the tradition that this stone "was the first solid thing created, and was placed by God amidst the as yet boundless fluid of the primeval waters.... And just as the body of the embryo receives its nourishment from the navel, so the

whole earth too receives the waters that nourish it from this Navel"; Patai, *Man and Temple: In Ancient Jewish Myth and Ritual* (London: Thomas Nelson and Sons, 1947), 85–86.

53. *The Fathers according to Rabbi Nathan*, 31:3.1, translation and explanation by Jacob Neusner (Atlanta: Scholars, 1986).

54. *The Fathers according to Rabbi Nathan*, 31:1–31:3: "'So you learn that one person is equivalent to all the works of creation.' A. R. Nehemiah says, 'How do we know that a single person is equivalent to all the works of creation? As it is said, This is the book of the generations of man, in the day that God created man, in the likeness of God he made him (Gen 5:1), and, elsewhere, These are the generations of heaven and earth when they were created (Gen 12:4). Just as in that latter passage we speak of creating and making, so in the former were creating and making. This teaches that the Holy One, blessed be he, showed him all the generations that were destined to come forth from him, as though they were standing and rejoicing before him.'"

55. *The Fathers according to Rabbi Nathan*, 31:3.1.

56. 2 Enoch 30:13.

57. *Sefer ha-temunah* (Lemberg, 1892), fols. 25a–b. Thus in Elliot Wolfson, "Mystical Rationalization of the Commandments in 'Sefer ha-Rimmon,'" *Hebrew Union College Annual* 59, no. 1988 (1988): 17n88.

58. Ibn Ezra, comment to Exodus 25:40, cited in Geoffrey W. Dennis, "Adam Kadmon," in *The Encyclopedia of Jewish Myth, Magic and Mysticism*, 2nd ed. (Woodbury, Minn.: Llewelyn Worldwide, 2016), 18.

59. Though in some texts the mapping is spiritual only, for others it is corporeal as well; Altmann, *Studies*, 16.

60. Chayyim Vital, *The Tree of Life*, vol. 1, *The Palace of Adam Kadmon*, trans. E. Collé and H. Collé (NP: CreateSpace, 2015), 12:2.7. The translators' footnotes include the mention of the dual translation. There are several transliterations of the author's name; I have used Chaim in my own text but have retained the spellings used by translators in citations.

61. Vital, *Tree of Life*, 13:2.8.

62. Vital, *Tree of Life*, 1:2.30. The number 248 appears in a number of Jewish descriptions, both mystical and rabbinic. It is the gematria of Abraham—that is, the number calculated via the numeric values assigned to the Hebrew letters of his name. These gematria have been in textual use since about the second century CE but are especially marked in Kabbalah.

63. *Zohar*, 3:128a. After careful considerations, the rabbis request, "May the prayer be accepted, that it may not be considered a sin to reveal this."

64. *Zohar* 3:128a.

65. *Zohar* 3:128a–b.

66. *Zohar* 3:128b.

67. Between *Zohar* 3:129a and 3:130b, we read of the "forehead of the skull," eyes without lids or brows, the nose concealed on every side, and the beard, its hairs as complex as those of the skull.

68. *Zohar* 3:128b–29a.

69. *The Secret Book of John: The Gnostic Gospel Annotated and Explained*, composed early second century CE, trans. and notes by Stevan Davies (Woodstock, Vt.: SkyLight Paths, 2005), 6:1–10 "The First Man/Who is the Image of the Invisible Spirit/Who is Barbelo/Who is Thought/And/Foreknowledge-Incorruptibility-Life Everlasting-Truth./ [These are an androgynous fivefold realm—therefore it is a realm of ten—of the Father]."

70. *Secret Book of John* 15, 15–25; "Demons," in Davies, *Secret Book of John*; "Angels," in *Secret Book of John*, trans. Marvin Meyer, in Meyer, *Nag Hammadi Scriptures*, and in *The Apocryphon of John*, trans. Frederick Wisse, in *The Nag Hammadi Library*, ed. James Robinson (New York: HarperCollins, 1990). Further notes here are to the Davies translation.

71. There may be a formless materiality in the microcosmic depictions in the *Zohar* as well. There formlessness belongs to the veil that first conceals the activity of creation, briefly floats unformed alone, and then is shaped to hint at divinity, which is itself without form. Though I do not know of a reading that identifies this veil with matter, it does act rather as matter does, formless without divinity and shaped by it in a way that can yield a revelation.

72. *Secret Book of John* 18, 10–15.

73. *Secret Book of John* 19, 1–5. As Stevan Davies tells us in his commentary, the careful assignment of demons to body parts and bodily functions probably had a medical significance, allowing their magic to be invoked or dispelled; Davies, *Secret Book of John* (location 1287), 15, 30–16,10. Seven demons govern the body as a whole; one each governs perceptions, reception, imagination, integration, and impulse; four govern heat, cold, dryness and damp; four govern the passions, 17, 30–18, 1.

74. Brakke, "Body as/at the Boundary of Gnosis," 204.

75. Brakke, "Body as/at the Boundary of Gnosis," 195.

76. Brakke, "Body as/at the Boundary of Gnosis," 205.

77. Cited and translated in Brakke, "Body as/at the Boundary of Gnosis," 205.

78. Altmann, *Studies*, 15, citing Isaac the Blind (ca. 1160–1235 CE), *Commentary on Sefer Yisirah* (ms.; Cincinnati: Hebrew Union College), fols. 35, 36.

79. Altmann, *Studies*, 14: "The *Sefer Bahir* (no. 55) sees in the seven (or six) limbs of the body of man an image of the seven (or six) lower Sefirot or six mystical days of creation, and applies to this analogy the verse, 'For in the image of God made He man' (Gen 9:6). This passage is reflected in *Tiqqune Ha-Zohar* (130b), where it is said, 'The limbs of man are all arranged in the order of the Beginning . . .'—that is, of the mystical days of creation which are identical with the six lower Sefirot—and man is therefore called a microcosm." The *Sefer Bahir* was probably written in the thirteenth century, and the *Tiqqune Ha-Zohar* is roughly dated 1100–1400 CE.

80. *Derech Eretz-Zutta* 9:13. Probably composed ca. 150–220, in Isaac Mayer Wise and Godfrey Taubenhaus, *New Edition of the Babylonian Talmud* (Boston: Talmud Society, 1918), 30.

81. *Secret Book of John* 14, 25–30.

NOTES TO PAGES 52–55

82. *Secret Book of John* 15, 1–5.
83. *Secret Book of John* 18, 10–15.
84. *Secret Book of John* 20, 25.
85. *Secret Book of John* 18–22.
86. *Secret Book of John* 20, 25–30.
87. *Secret Book of John* 23, 5–10.
88. *Secret Book of John* 4, 25–35.
89. See Plotinus, *Ennead* 1.6.9, in *The Enneads*, trans. Stephen MacKenna (London: Penguin, 1991). Plato says that the eye is the most sun-like organ; *Republic*, trans. Robin Waterfield (Oxford: Oxford University Press, 2008), 508b.
90. *Secret Book of John* 4, 1–10.
91. *Leviticus Rabbah* 20, 2.
92. Heraclitus, fragment B3, in Patricia Curd, ed., trans., *A Presocratics Reader* (Indianapolis: Hackett, 2011), 50.
93. *Bava Batra* 58a, composed ca. 450–550 CE, trans. Adin Even-Israel Steinsaltz, accessed March 2021, at https://www.sefaria.org/Bava_Batra.58a?lang=bi.
94. Altmann, *Studies*, 157; cf. Altmann, "Gnostic Themes in Rabbinic Cosmology," in *Essays in Honour of the Very Rev. Dr. J. H. Hertz*, ed. I. Epstein, E. Levine, and C. Roth (London: E. Goldston, 1942), 28–52.
95. Altmann, "Gnostic Themes," 30, citing *Genesis Rabbah* 12:6.
96. Vital, *Tree of Life*, 1:1.1.8, p. 11.
97. *Secret Book of John* 29, 1–15.
98. *The Gospel of Philip* 70, 5–9: "The powers cannot see those who have put on the perfect light, and they cannot seize them. One puts on the light in the mystery of union."
99. The *Untitled Treatise* brings the light together with divinity by making the savior itself into a light-garment, perhaps returning us by an indirect route to light's protective character. Brakke describes the savior here as "a garment of the entireties, as our Human Being wears the glory of the aeons as a garment"; Brakke, "Body as/at the Boundary of Gnosis," 211, citing *Untitled Treatise* 91.35. Chapters 12–13 of the *Treatise* describe a garment given to Solomon: "The first monad furthermore sent him an ineffable garment which was all light and all life and all resurrection, and all love and all hope and all faith and all wisdom, and all gnosis, and all truth, and all peace, and all-visible. . . . And the All is in it, and also all found themselves in it, and knew themselves in it. And it (the monad) gave light to them all with its ineffable light"; *Untitled Treatise*, in *The Books of Jeu and the Untitled Text in Bruce Codex*, ed. Carl Schmidt, trans. Violet McDermot (Leiden: Brill, 1978), 250–51.
100. "The Holy One, blessed be He, wrapped Himself [in light] as in a robe and irradiated with the lustre of His majesty the whole world from one end to the other"; *Genesis Rabbah* 3:4. The same point appears at 1:6. The passage makes reference to Psalm 104:1–2, "You are clothed with honor and majesty, wrapped in light as with a garment."
101. *Genesis Rabbah* 20:12.

102. Genesis 3:21: "The Lord God made garments of skin for Adam and his wife, and clothed them."

103. *Genesis Rabbah* 12:6.

104. Isaac Luria, *Collected Works on Kabbalah*, as promulgated by his student, Rabbi Hayyim Vital (1543–1620): *Sefer ha-Likkutim* (Jerusalem, 1988), 15:28, cited and translated in Stanley Schneider and Morgan Seelenfreund, "Kotnot Or (Genesis 3:21): Skin, Leather, Light, or Blind?," *Jewish Bible Quarterly* 40, no. 2 (2012): 118.

105. Genesis 3:7–8.

106. *Targum Pseudo-Jonathan*, on Genesis 3:9.

107. Genesis 3:8–24.

108. Where the human is read as microcosm, human diminution must damage the cosmos, too. For Isaac Luria, the error of the first Adam undoes the drawing-together of the levels of the world, the task he was intended to accomplish, which entailed the gathering together of divine light; see Lawrence Fine, "Tikkun: A Lurianic Motif in Contemporary Jewish Thought," in *From Ancient Israel to Modern Judaism: Intellect in Quest of Understanding—Essays in Honor of Marvin Fox*, ed. Jacob Neusner, Ernest S. Frerichs, and Nahum M. Sarna, vol. 4 (Providence, R.I.: Brown Judaic Studies, 1989).

109. *Zohar* 1:131b.

110. In a sermon on Sirach, the medieval Dominican Meister Eckhart says, "The eye with which I see God is the same with which God sees me. My eye and God's eye is one eye, and one sight, and one knowledge, and one love"; German sermon on Ecclesiasticus 24:30, in *Meister Eckhart, Selected Writings*, ed. and trans. Oliver Davies (New York: Penguin, 1994), 179.

111. This is less true of capital-O Orthodoxy; Eastern Christianity has long been more comfortable with theosis (divinization) than the Western version has.

112. Marvin Meyer, "Gnosticism, Gnostics, and the Gnostic Bible," introduction to *The Gnostic Bible*, ed. Willis Barnstone and Marvin W. Meyer (Boston: Shambhala, 2003), 8. Here Meyer is commenting on a passage from *The Book of Thomas the Contender*: "Examine yourself and understand who you are, how you exist, and how you will come to be. . . . It is not fitting for you to be ignorant of yourself"; *The Book of Thomas*, trans. Marvin Meyer, in Meyer, *The Nag Hammadi Scriptures* (New York: HarperCollins, 2008), 138, 4–15. A similar idea emerges in the Sethian *Allogenes*: "If you [seek with a perfect] seeking, [then] you shall know [the Good that is] in you; then [you will know yourself] as well"; *Allogenes*, trans. John D. Turner and Orval S. Wintermute, in Robinson, *Nag Hammadi Library*, 56, 15–20.

113. *Gospel of Philip*, trans. Marvin Meyer, in Meyer, *Nag Hammadi Scriptures*, 57, 22–28.

114. See particularly Maurice Merleau-Ponty, *The Visible and the Invisible*, trans. Alphonso Lingis (Evanston, Ill.: Northwestern University Press, 1968), and Jacques Lacan's seminars on the gaze in *The Four Fundamental Concepts of Psychoanalysis*, ed. Jacques-Alain Miller (New York: W. W. Norton, 1998), chapters 6–9.

115. *Gospel of Philip*, 86, 15–19.

116. Betty Rojtman, *Black Fire on White Fire: An Essay on Jewish Hermeneutics, from Midrash to Kabbalah*, trans. Steven Rendall (Berkeley: University of California Press, 1998), 151.

117. Karen King notes that the *Secret Book of John* is rather convoluted, in part as a result of "the intertextual reading of Platonic cosmology, Genesis, and Wisdom literature." The result is a multiplication: "The 'double' reading of Genesis and Wisdom on a dualist Platonizing framework produces two creator gods, two wisdoms, two Eves, and so on, one belonging to perfection, one to parody. Wisdom is the glue that holds all of these together. Wisdom is Pronoia, Sophia, and Eve (as well as Christ, Epinoia, and Yaldabaoth); she is the creator above (Pronoia) and below (Yaldabaoth); she is the savior above (Pronoia, Christ) and below (Epinoia, Christ)"; King, *The Secret Revelation of John* (Cambridge, Mass.: Harvard University Press, 2009), 234.

3. Limitless Bounding: The Valentinian Body of Christ

1. Moses Maimonides, *Mishnah Torah: Kings and Wars*, composed ca. 1176–78 CE, chap. 11, trans. Reuven Brauner, 2012; accessed March 27, 2021, at https://www.sefaria.org/Mishneh_Torah%2C_Kings_and_Wars.11.1?ven=Laws_of_Kings_and_Wars._trans._Reuven_Brauner,_2012&lang=bi. In all notes, where the URL differs between the footnote and bibliographic references, the footnote reference is to the cited point in the text and the bibliographic reference is to the starting point of the text.

2. Moses Maimonides, commentary on *Mishnah Sanhedrin*, composed ca. 1158–68 CE, chap. 10, trans. Sefaria Community Translation, accessed March 27, 2021, at https://www.sefaria.org/Rambam_on_Mishnah_Sanhedrin?lang=bi.

3. Few among the very rare exceptions to this masculinity are notable. One, being Christian, might merit mention: Mother Ann Lee founded the Christian sect popularly called the Shakers; she was regarded by her followers as a second messianic figure, parallel to Christ.

4. David Brakke, "The Body as/at the Boundary of Gnosis," *Journal of Early Christian Studies* 17, no. 2 (Summer 2009): 207, citing Genesis 1:26, Ezekiel 1:26. Michael Anthony Knibb also notes several possible references to a divine messiah in the Hebrew Bible (Psalm 2:7), the Qumran texts 4Q246 and 1Q28a, and five Jewish pseudepigraphic texts: the Psalms of Solomon 17 and 18, 1 Enoch, 4 Ezra, 2 Baruch, and the Testaments of the Twelve Patriarchs. The pseudepigrapha, as he notes, may show some Christian influence; Knibb, *Essays on the Book of Enoch and Other Early Jewish Texts and Traditions* (Leiden: Brill, 2009), 307–11.

5. Hebrews 1:2–3. Unless otherwise noted, all biblical translations are from the New Revised Standard Bible.

6. Scholars disagree about the exact sources of the hymn in Colossians but do agree that it is older than the rest of the letter and has common ground with descriptions of Wisdom. Matthew Gordley reviews many of the arguments in *The*

Colossian Hymn in Context: An Exegesis in Light of Jewish and Greco-Roman Hymnic and Epistolary Conventions (Tübingen: Mohr Siebeck, 2007).

7. Colossians 1:15–17.

8. John 1:1–4.

9. Augustine, *On the Literal Interpretation of Genesis: An Unfinished Book*, trans. Roland J. Teske, 3.6, in *Augustine on Genesis: Two Books on Genesis against the Manichees and the Literal Interpretation of Genesis, an Unfinished Book*, ed. Roland J. Teske (Washington, D.C.: The Catholic University of America Press, 1991). The translation necessarily obscures some of the entanglements; "In the beginning" is "In principio" in Augustine's text and in the Latin biblical translations he would have known. The Greek *logos* can be rendered both "principle" and "word."

10. Augustine, *On the Literal Meaning of Genesis*, trans. John Hammond Taylor (Mahwah, N.J.: Paulist Press, 1982), 4.9, 10.20.

11. John 1:14.

12. Augustine, *Confessions*, trans. Henry Chadwick (Oxford: Oxford University Press, 1991), 9.19–21.

13. This idea is called Docetism and was rejected by the earliest ecumenical councils. It is often attributed to groups called Gnostic, not always with much supporting evidence.

14. 1 John 2:2.

15. More exactly, in mainstream Western Christianity, in an orthodoxy developed by Augustine, humans cannot lose their tendency to sin, and any redemption can come about only through divine grace. In Eastern Christianity, a more generous interpretation holds that humans can, with effort and divine help, overcome even the backsliding tendency to sin again, and can attain purity in their lifetimes—though such attainment is rare and difficult. The possibility is perhaps clearest as an accomplishment at which monks might aim, as in John Cassian, *The Conferences*, trans. Boniface Ramsey (Mahwah, N.J.: Paulist Press, 1997), esp. Conference 19.

16. *The Gospel of Truth*, trans. Marvin Meyer, in *The Nag Hammadi Scriptures*, ed. Marvin Meyer (New York: HarperCollins, 2008), 16, 31–35.

17. *Gospel of Truth* 18, 11–15.

18. See, as a very incomplete set of instances, the *Gospel of Truth* and the *Treatise on the Resurrection* (also called the *Epistle to Rheginus*), trans. Marvin Meyer, in Meyer, *Nag Hammadi Scriptures*; and Irenaeus, *Against Heresies*, in *Ante-Nicene Fathers*, vol. 1, ed. Alexander Roberts, James Donaldson, and A. Cleveland Coxe (Buffalo, N.Y.: Christian Literature, 1885), 1.15.3, 3.16.1.

19. Brakke, "Body as/at the Boundary of Gnosis," 197.

20. Irenaeus, *Against Heresies* 1.3.3.

21. Patricia Cox Miller, "'Plenty Sleeps There': The Myth of Eros and Psyche in Plotinus and Gnosticism," in *Neoplatonism and Gnosticism*, ed. Richard T. Wallis and Jay Bregman (Albany: State University of New York Press, 1992), 226: "The figures who compose the pleroma, like the 'real beings' of Plotinus's realm of 'Nous,' are the metaphors of divine reality; they are the collection of signposts that dot the

'road and the travelling' of human attempts to express in language the profound mystery at the heart of things."

22. Aristotle, *Metaphysics*, trans. Hugh Tredennick (London: William Heinemann, 1989), 986a.

23. Plato, *Republic*, trans. Robin Waterfield (Oxford: Oxford University Press, 2008), 509d–11e.

24. Plato, *Phaedrus*, trans. Alexander Nehamas and Paul Woodruff (Indianapolis: Hackett, 1995), 265d–66a.

25. Plato, *Symposium* 189c–93e.

26. Plato, *Statesman*, trans. Eva Brann, Peter Kalkavage, and Eric Salem (Indianapolis: Hackett, 2012), 262a–b.

27. See especially Plato, *Statesman* 262c–63b.

28. Plato, *Theaetetus*, trans. John McDowell (Oxford: Oxford University Press, 2014), 147c–48b.

29. Plato, *Theaetetus* 143e–44d.

30. Plato, *Statesman* 257d.

31. Plato, *Theaetetus* 210c.

32. Plato, *Parmenides*, trans. Harold N. Fowler (London: William Heinemann, 1925), 129c-d.

33. Plato, *Statesman* 263b.

34. Plato, *Parmenides* 131a–33b.

35. This part of the argument is fairly well known on its own as the "third man argument"; Plato, *Parmenides* 132a–b.

36. Sarah Pessin, *Ibn Gabirol's Theology of Desire: Matter and Method in Jewish Medieval Neoplatonism* (Cambridge: Cambridge University Press, 2013), 166.

37. Plato, *Parmenides* 137d.

38. Kenneth Sayre, *Plato's Late Ontology: A Riddle Revised* (Princeton, N.J.: Princeton University Press, 1983), 120. The passage from Aristotle's *Metaphysics* reads, "In treating the One as a substance instead of a predicate of some other entity, his teaching resembles that of the Pythagoreans, and also agrees with it in stating that the numbers are the causes of Being in everything else; but it is peculiar to him to posit a duality instead of the single Unlimited, and to make the Unlimited consist of the 'Great and Small'"; Aristotle, *Metaphysics*, 987b25–27.

39. Aristotle, *Metaphysics* 987b.

40. Important voices here include Sayre, *Plato's Late Ontology*, and Mitchell Miller, "The Choice between the Dialogues and the 'Unwritten Teachings': A Scylla and Charybdis for the Interpreter?," in *The Third Way: New Directions in Platonic Studies*, ed. Francisco Gonzalez (Lanham, Md.: Roman and Littlefield, 1995).

41. See Sayre, *Plato's Late Ontology*, and Miller, "Choice." Konrad Gaiser also cites Hans Joachim Kramer, *Plato and the Foundations of Metaphysics: A Work on the Theory of the Principles and Unwritten Doctrines of Plato with a Collection of the Fundamental Documents* (Albany: State University of New York Press, 1990), 123f; Konrad Gaiser, "Plato's Enigmatic Lecture 'On the Good,'" *Phronesis* 25, no. 1

(1980): 29n9. See also Joel Kalvesmaki, *The Theology of Arithmetic: Number Symbolism in Platonism and Early Christianity* (Washington, D.C.: Center for Hellenic Studies, 2013), 45. Kalvesmaki cites Aristotle, *Metaphysics* 987b29–88a1, 1091a13–29; Walter Burkert, *Lore and Science in Ancient Pythagoreanism* (Cambridge, Mass.: Harvard University Press, 1972), 22; Julia Annas, *Aristotle's Metaphysics Books M and N* (Oxford: Oxford University Press, 1976), 43. For a vivid application of this definition of number, see Plato, *Parmenides* 143c–44a.

42. Simplicius includes a summary of a lost commentary by Porphyry on Aristotle's categories. Sayre points out several other ancient texts that identified the one's paired dyad as the great and small, or the dyad as the indefinite and unlimited, in *Plato's Late Ontology*, esp. 149–51.

43. See esp. Christopher Riedweg, *Pythagoras: His Life, Teaching, and Influence*, trans. Steven Rendall (Ithaca, N.Y.: Cornell University Press, 2005), 118.

44. Aristotle, *Metaphysics* 988a10–15, and Theophrastus, *On First Principles: Known as His Metaphysics*, trans. D. Gutas (Leiden: Brill, 2010), 6b10–15.

45. John Pepple, "The Unwritten Doctrines: Plato's Answer to Speusippus," 1997, sec. 7, accessed March 27, 2021, at http://personal.kenyon.edu/pepplej/#section%20VII.

46. Lenn E. Goodman, introduction to *Neoplatonism and Jewish Thought*, ed. Lenn E. Goodman (Albany: State University of New York Press, 1992), 6, citing Aristoxenus, *Elementa Harmonica* 2.30–31.

47. W. K. C. Guthrie, *A History of Greek Philosophy*, vol. 5, *Later Plato and the Academy* (Cambridge: Cambridge University Press, 1978), 424–26, cited in Gaiser, "Plato's Enigmatic Lecture," 6.

48. Konrad Gaiser, "Plato's Enigmatic Lecture," 5.

49. Plato, *Philebus*, trans. Dorothea Frede (Indianapolis: Hackett, 1993), 16c.

50. Plato, *Philebus* 16b; Sayre, *Plato's Late Ontology*, 118.

51. See, e.g., *Philebus* 23d–e, 25A; 23e–24a.

52. Plato, *Philebus* 65b.

53. John D. Turner, *Sethian Gnosticism and the Platonic Tradition* (Peeters, Belgium: Le Presses de l'Université Laval, 2006), 315–16, citing Aristotle, *Metaphysics* I.6 987b32–3; cf. Plato, *Timaeus*, trans. W. R. M. Lamb (London: William Heinemann, 1925), 50C. In the *Philebus*, this division introduces proportion to the continuous Unlimited and thus makes balance and harmony possible.

54. Joel Kalvesmaki, *Theology of Arithmetic*, 45. Kalvesmaki cites Aristotle, *Metaphysics* 1088a4–8, and a similar point in Euclid's *Elements*, book 7, defs. 1–2: "A monad is that by which each existent thing is called one. Number is a multitude composed of monads."

55. Kalvesmaki, *Theology of Arithmetic*, 45–46.

56. Kalvesmaki, *Theology of Arithmetic*, 46.

57. Plotinus, *Plotinus: The Enneads*, trans. Stephen MacKenna (New York: Penguin, 1991), 5.1.1.

58. Plotinus, *Enneads* 6.7.17.

59. The latter will be the focus of this chapter, but a particularly interesting version of the former can be found in the medieval Jewish philosopher Solomon ibn Gabirol, whose ideas are influenced by Islamic Aristotelianism as well as by Platonism. For a good summary, see John M. Dillon, "Solomon Ibn Gabirol's Doctrine of Intelligible Matter," in Goodman, *Neoplatonism and Jewish Thought*, 43–59.

60. Turner, *Sethian Gnosticism*, 355.

61. Elliot R. Wolfson, *Venturing Beyond: Law and Morality in Kabbalistic Mysticism* (Oxford: Oxford University Press, 2006), 205n69. For a discussion of the neo-Platonic axiom that simplicity can only give rise to simplicity, Wolfson refers to Arthur Hyman, "From What Is One and Simple Only What Is One and Simple Can Come to Be," in Goodman, *Neoplatonism and Jewish Thought*, 111–35.

62. Dillon, "Solomon Ibn Gabirol's Doctrine of Intelligible Matter," 44.

63. Turner, *Sethian Gnosticism*, 365–66.

64. Turner, *Sethian Gnosticism*, 372.

65. Wolfson, *Venturing Beyond*, 205, my emphasis. Glossing *Zohar* 3:80b: "When the blessed Holy One unveiled the world and sought to reveal the deep from within the hidden and light from within darkness, they were intermingled—for out of darkness issued light, and out of the hidden was revealed the deep, one issuing from the other. Out of good issues evil, out of Compassion issues Judgment. All is interwoven . . . all suspended in oneness"; *Zohar*, trans. Daniel C. Matt (Stanford, Calif.: Stanford University Press, 2003).

66. Aristotle mentions the Pythagorean table of opposites, which lists valued and disvalued qualities in two columns, in *Metaphysics* 986b.

67. Turner, *Sethian Gnosticism*, 362.

68. Kalvesmaki, *Theology of Arithmetic*, 46, 55.

69. Hippolytus, *Refutation of All Heresies*, trans. J. H. MacMahon, in *Ante-Nicene Fathers*, vol. 5, ed. Alexander Roberts, James Donaldson, and A. Cleveland Coxe (Buffalo, N.Y.: Christian Literature, 1886), 6.29; Kalvesmaki, *Theology of Arithmetic*, 49f.

70. Irenaeus, *Against Heresies* 1.10.5–1.11.1

71. Kalvesmaki, *Theology of Arithmetic*, 36n24: "Irenaeus claims that the Valentinians (1) have a common false doctrine and (2) constantly disagree with each other, each variation being a lie"; "Tertullian accuses the Valentinians of vacillating between monism and dualism" (54).

72. Kalvesmaki, *Theology of Arithmetic*, 53: "Another Nag Hammadi text, *A Valentinian Exposition*, also offers a form of dyadic Valentinianism with monadic elements. Previous commentators and editors have treated the text as being primarily monadic, but there are no good reasons to do so. . . . The Father has a companion, Silence, and they form a dyad. Yet the Father exists monadically, a special solitude he enjoys within his dyadic condition." For a general description of the text, see Birger Pearson, *Ancient Gnosticism: Traditions and Literature* (Minneapolis: Fortress, 2007), 182–83.

73. *The Tripartite Tractate*, trans. Harold W. Attridge and Dieter Mueller, in *The Nag Hammadi Library*, ed. James M. Robinson (New York: HarperCollins, 1990), 1 51, 5–15.

74. A *Valentinian Exposition*, trans. Einar Thomassen and Marvin Meyer, in Meyer, *The Nag Hammadi Scriptures*, 25, 12–13. The numbering in this edition does not always allow exact citation; where I have been uncertain, I have indicated approximate lines.

75. Augustine, *On the Literal Interpretation of Genesis: An Unfinished Book*, 3.6.

76. Cf. *Allogenes*, a Sethian text used by the Valentinian school: "O Allogenes, behold your blessedness how it silently abides, by which you know your proper self and, seeking yourself, withdraw to the Vitality that you will see moving . . . if you wish to stand, withdraw to the Existence, and you will find it standing and at rest after the likeness of the One who is truly at rest and embraces all these silently and inactively"; *Allogenes*, trans. John D. Turner and Orval S. Wintermute, in Robinson, *The Nag Hammadi Library*, 59, 10–20.

77. *Gospel of Truth* 17, 5–9.

78. *Gospel of Truth* 17, 4.

79. *Valentinian Exposition* 22, 19–25.

80. *Valentinian Exposition* 22, 19–25.

81. *Valentinian Exposition* 22, 25–30.

82. *Zostrianos*, trans. John D. Turner, in Meyer, *Nag Hammadi Scriptures*, 118, 1–10.

83. *Allogenes* 60, 20–29.

84. *The Secret Book of John: The Gnostic Gospel Annotated and Explained*, trans. and annotated Stevan Davies (Woodstock Vt.: SkyLight Paths, 2005), 3, 5.

85. Kalvesmaki, *Theology of Arithmetic*, 52, citing Irenaeus, *Against Heresies* 1.29.1–4.

86. Kalvesmaki, *Theology of Arithmetic*, 52. As he notes, "Systems where the second principle envelops the first appear frequently in Pythagorean texts of the period." He cites Einar Thomassen, *The Spiritual Seed: The Church of the "Valentinians"* (Leiden: Brill, 2006), 293–94.

87. James Moffatt, "Pistis Sophia," in *Encyclopaedia of Religion and Ethics*, ed. James Hastings (New York: Charles Scribner's Sons, 1919), 10:46. The text proposes as a possible etymology of *Barbelo*, "The supreme limit, . . . from the Indian *vela*, 'limit,'—a suggestion made by Julias Grill (*Untersuchungen über die Entstehung des vierten Evangeliums* [Tübingen, 1902], 395–97), who connected it with the Valentinian *Horos*."

88. Kalvesmaki, *Theology of Arithmetic*, 53.

89. *Valentinian Exposition* 22.31. The *Untitled Tractate* that Brakke discusses makes a similar claim: "The Father whose spoken word penetrates both upper and lower regions is also a spring that pours forth silence"; 226.20, cited in Brakke, "Body as/at the Boundary of Gnosis," 212.

90. *Valentinian Exposition* 23.32–24.5.

91. *Valentinian Exposition* 24, ca. 19–35.

92. The author of the *Valentinian Exposition* "call[s] him, with reference to the Thought, the Only One." This Thought seems to be the image of the first thought, Truth, since Truth "brought forth the Only One and the [Boundary]"; *Valentinian Exposition* 24, 40; 25.

NOTES TO PAGES 74-76   161

93. *Valentinian Exposition* 25, 22–25. Final ellipsis mine.
94. *Valentinian Exposition* 26, ca 25–30.
95. *Valentinian Exposition* 26, 22.
96. *Valentinian Exposition* 25, 30.
97. *Valentinian Exposition* 25, 30.
98. Brakke, "Body as/at the Boundary of Gnosis," 209.
99. See Anne Carson, *Eros the Bittersweet: An Essay* (Princeton, N.J.: Princeton University Press, 1986), *passim*.
100. Clement of Alexandria, *Excerpta Ex Theodoto of Clement of Alexandria*, composed ca. 175–200 CE, trans. Robert Pierce Casey (London: Christophers, 1934), 26. Theodotus is discussing John 10:7–9, in which Jesus refers to himself as "a gate for the sheep."
101. *Tripartite Tractate* 71, 14–19.
102. Irenaeus, *Against Heresies* 1.2.1.
103. Tertullian, *Against the Valentinians*, chap. 9, trans. Mark T. Riley, 1971, accessed March 27, 2021, at http://tertullian.org/articles/riley_adv_val/riley_00_index.htm.
104. Irenaeus, *Against Heresies* 1.2.2.
105. Tertullian, *Against the Valentinians*, chap. 9.
106. Irenaeus, *Against Heresies* 1.2.2.
107. Tertullian, *Against the Valentinians*, chap. 9.
108. Cf. Brakke, "Body as/at the Boundary of Gnosis," 209: "The Valentinian insight that separation or differentiation from God is not only tragic but also salutary took mythological form in the character of Horos, the Limit or Boundary. According to Irenaeus, this character appears already in Valentinus's own myth in two forms, an inner boundary that separates the deep from the rest of the fullness and an outer boundary that separates the fullness from the lower mother (and thus from materiality in general)."

There is another version of Valentinian cosmology, which is closer to some other ideas collected under the heading of Gnosticism, in which Sophia's error is not this quest for knowledge but the desire to imitate the Father by creating on her own. Without pretending that this tradition is the least bit unimportant, I want here to attempt a reading only of the knowledge-seeking story; see David Brons, "Valentinian Theology," accessed March 27, 2021, at http://gnosis.org/library/valentinus/Valentinian_Theology.htm.

109. Irenaeus, *Against Heresies* 1.2.2.
110. Brons, "Valentinian Theology." Brons refers to Irenaeus, *Against Heresies* 1.2.3, and *Valentinian Exposition* 34, 25–31.
111. Brons, "Valentinian Theology," with reference to Irenaeus, *Against Heresies* 1.2.4, and Hippolytus, *Refutation* 31.5.
112. Irenaeus, *Against Heresies* 1.4.1.
113. Irenaeus, *Against Heresies* 1.4.1.
114. Irenaeus, *Against Heresies* 1.4.1–2. Irenaeus continues by declaring all of this absurd and stating firmly that it is right that the Valentinians kept their teachings

from the public, as they were too embarrassingly bad to be widely proclaimed; Irenaeus, *Against Heresies* 1.4.1–2; Hippolytus, *Refutation* 6.27, *Tripartite Tractate* 81, 22–83, 33; all referred to in Brons, "Valentinian Theology."

115. *Gospel of Truth* 17, 5–10.

116. Tertullian, *Against the Valentinians*, chap. 16.

117. Hippolytus, *Refutation* 6.27.

118. John P. Kenny, "The Platonism of the Tripartite Tractate," in *Neoplatonism and Gnosticism*, ed. Richard T. Wallis and Jay Bregman (Albany: State University of New York Press, 1992), 194.

119. Brakke, "Body as/at the Boundary of Gnosis," 211.

120. *Valentinian Exposition* 27, 20–21.

121. Irenaeus, *Against Heresies* 14. Irenaeus's version of Valentinianism here makes the proximity to Paul especially clear. In the hymn from the Letter to the Colossians to which this chapter earlier referred, we read, "He is the image of the invisible God, the firstborn of all creation; for in him all things in heaven and on earth were created, things visible and invisible, whether thrones or dominions or rulers or powers—all things have been created through him and for him" (Col. 1:15–16).

122. *Valentinian Exposition* 35, ca. 1–5.

123. *Valentinian Exposition* 35, ca. 10–25.

124. *Valentinian Exposition* 35, 25–36, 1.

125. *Valentinian Exposition* 36, 1–18.

126. *Tripartite Tractate* 90, 30–35.

127. *Valentinian Exposition* 36, 15–20.

128. *Tripartite Tractate* 104, 25.

129. Until recently, scholars have almost exclusively emphasized the negative claim that an image is inferior to the original. That the more positive emphasis is possible, however, also has scholarly support; see especially Christoph Markschies, *Gnosis: An Introduction*, trans. John Bowen (London: T & T Clark, 2003), 89–94. Dunderberg calls Markschies's argument "compelling," noting that it is supported not only by Valentinus's Fragment 8, which Markschies cites, but also by Fragment 5 and Valentinus's hymn "Summer Harvest," and that some of the same metaphors that we find in "Summer Harvest" are at work in the *Gospel of Truth* (34–35) and the *Gospel of Philip*, in Meyer, *Nag Hammadi Scriptures* (109, 7–8); Ismo Dunderberg, "Valentinus and His School," *Revista Catalana de Teologia* 37, no. 1 (2012): 139–41.

130. Origen, *On First Principles*, trans. G. W. Butterworth (Gloucester, Mass.: Peter Smith, 1973), 2.9.1, Latin: "In the beginning God made as large a number of rational and intelligent beings . . . as he foresaw would be sufficient. It is certain that he made them according to some definite number fore-ordained by himself, for we must not suppose, as some would, that there is no end of created beings, since where there is no end there can be neither comprehension nor limitation."

131. *Tripartite Tractate* 122, 32–123, 22.

132. Colossians 1:14, 1 Corinthians 3:17, Ephesians 5:30; cf. Tertullian, *On Modesty*, trans. S. Thelwall, in *Ante-Nicene Fathers*, vol. 4, chap. 16; Irenaeus,

Against Heresies 5.2; Augustine, "Sermon 272—On the Day of Pentecost," in *The Works of Saint Augustine: A Translation for the 21st Century; Sermons 230–272B*, vol. 3, part 7, *On the Liturgical Seasons*, trans. Edmund Hill, O.P. (Hyde Park N.Y.: New City, 1990), and *City of God*, trans. Marcus Dods, ed. Philip Schaff, *Nicene and Post-Nicene Fathers 1*, vol. 2 (Grand Rapids, Mich.: William B. Eerdmans, 1887), 10.6; Clement of Alexandria, *Excerpta Ex Theodoto* frag., 42; Brons, "Christ and the Church," accessed March 27, 2021, http://gnosis.org/library/valentinus/Christ_and_Church.htm. Brons refers as well to the canonical books Romans 12:5; 1 Corinthians 12:12–13; Ephesians 4:16; and Colossians 1:18, 2:19. He also refers to Irenaeus, *Against Heresies* 1.7.2 and the *Tripartite Tractate* 122, 12–17.

133. Brons, "Christ and the Church," with reference to the *Valentinian Exposition* 30, 28–30.

134. Theodotus, *Excerpta ad Theodoto* 42.3.

135. David Brons, "Christ and the Church," my ellipsis.

136. David Brons, "Christ and the Church," citing Herakleon, Fragment 39.

137. Tertullian, *Prescription against Heretics*, in Ante-Nicene Fathers, vol. 3, ed. Alexander Roberts, James Donaldson, and A. Cleveland Coxe (Buffalo, N.Y.: Christian Literature, 1885), chap. 41.

138. *Gospel of Truth* 19, 17.

139. *Gospel of Truth* 30, 23.

140. *Gospel of Truth* 19, 17–34.

141. Plato, *Apology*, 21B–23A, 20D–20E, and 23A.

142. Brakke, "Body as/at the Boundary of Gnosis," 214.

143. *Gospel of Truth* 19, 34–20, 5.

144. *Gospel of Truth* 21, 1–7.

145. *Gospel of Truth* 21, 4–7, 8.

146. *Gospel of Truth* 19, 34–ca. 20, 10.

147. Here too there is an overlap with later canonical Christian scriptures, in the declaration from John 1:14, "And the Word was made flesh, and lived among us."

148. Irenaeus, *Against Heresies* 1.3.4–6. Irenaeus notes that the Valentinians also cite canonical Christian scriptures as supportive evidence in their descriptions of Limit, including Luke 14:27 and 3:17, Matthew 10:21 and 10:34, and 1 Corinthians 1:18. He writes that they "maintain that John" had Limit in mind "when he said, 'The fan is in His hand, and He will thoroughly purge the floor, and will gather the wheat into His garner; but the chaff He will burn with fire unquenchable.' By this declaration He set forth the faculty of Horos. For that fan they explain to be the cross (Stauros), which consumes, no doubt, all material objects, as fire does chaff, but it purifies all them that are saved, as a fan does wheat"; *Against Heresies* 1.3.5.

149. Tertullian, *Against the Valentinians*, chap. 9.

150. Tertullian, *Against the Valentinians*, chap. 27.

151. Clement of Alexandria, *Excerpta ex Theodoto* 42. Theodotus has in mind a curious passage from the Letter to the Ephesians: "We are to grow up in every way into him who is the head, into Christ, from whom the whole body, joined and knit

together by every joint with which it is supplied, when each part is working properly, makes bodily growth and upbuilds itself in love" (Eph 4:15–16).

152. Genesis 1:5.

153. Brons, "Valentinian Theology": "The Savior is associated with a retinue of angels who are the prototypes of the spiritual element present in every Christian. Like rays of the sun, they are not distinct or self-sufficient individuals. Rather, they represent the dynamic richness of Jesus." Brons cites Irenaeus, *Against Heresies* 1.2.6, and Clement of Alexandria, *Excerpta ex Theodoto* 39–40.

154. Clement of Alexandria, *Excerpta ex Theodoto* 22.

155. *Gospel of Truth* 22, 30–23, 17.

156. Miller, "'Plenty Sleeps There,'" 11.

157. Patricia Cox Miller, "'Words with an Alien Voice': Gnostics, Scripture, and Canon," *Journal of the American Academy of Religion* 57, no. 3 (1989): 460.

158. *Gospel of Truth* 38, 6.

159. *Gospel of Truth* 38, ca. 10.

160. *Gospel of Philip* 56, 3–13.

161. Irenaeus, *Against Heresies* 1.14.1.

162. See Brons, "The Name and Naming in Valentinianism," accessed March 27, 2021, http://gnosis.org/library/valentinus/Name_Naming.htm, citing *Gospel of Truth* 22: "This astonishing idea has its root in the notion that the emanation of the Name by the Father was a process of self-limitation. Valentinus himself admits that it is a surprising idea, 'It was quite amazing that they were in the Father without being acquainted with him and that they alone were able to emanate, inasmuch as they were not able to perceive and recognize the one in whom they were.' The Aeons can be thought of as unintegrated aspects of the Son's overall personality who are unaware of the Name even while they form part of it."

163. *Gospel of Truth* 24, 29–32.

164. Edward Moore, "Gnosticism," in *Internet Encyclopedia of Philosophy*, accessed March 27, 2021, at http://www.iep.utm.edu/gnostic/. Parenthesis mine.

165. Kenny, "Platonism of the Tripartite Tractate," 202.

166. Christoph Bandt, "Introduction to Fractals," in *Fractals, Wavelets, and Their Applications*, ed. Christoph Bandt (New York: Springer International, 2014), 5.

167. Though there are others, the Koch snowflake is probably the best known of these infinite limits and is well explained at http://mathworld.wolfram.com/KochSnowflake.html.

168. Benoit Mandelbrot, *The Fractal Geometry of Nature* (New York: W. H. Freeman, 1982), 147.

169. Mandelbrot, *Fractal Geometry*, 149.

170. Orly H. Shenker, "Fractal Geometry Is Not the Geometry of Nature," *Studies in the History and Philosophy of Science* 25, no. 6 (1994): 969.

171. Natural trees do not grow so neatly; fractal trees are geometric abstractions—though abstractions that are visually recognizable as trees as soon as we see them.

172. Kalvesmaki, *Theology of Arithmetic*, 46.

173. Einar Thomassen, "The Structure of the Transcendent World in the Tripartite Tractate (NHC I, 5)," *Vigilae Christianae* 34, no. 4 (1980): 369.

174. Thomassen, "Structure of the Transcendent World," 370. He continues, "This is apparently what is meant by 'the Limit' which causes them to be silent about the Father but to speak of their desire to know him (*Tripartite Tractate* 75, 13–17, cf. 72, 25–27)."

175. Moore, "Gnosticism."

176. Miller, "'Plenty Sleeps There,'" 227, citing Irenaeus, *Against Heresies* 1.11.4; *On the Origin of the World*, trans. Marvin Meyer, in Meyer, *Nag Hammadi Scriptures*, 5, 109.

177. Hippolytus, *Refutation* 6.24. The productivity of love runs across Platonic, neo-Pythagorean, and, in the Middle Ages, Pseudo-Empedoclean philosophies as they are taken up across the Abrahamic faiths. It is strongly present in Ibn Gabirol's work; see Pessin, *Ibn Gabirol's Theology of Desire*, 170. For a summary of Pseudo-Empedoclean philosophy in Islamic thought, see Lenn Goodman, "Ibn Masarrah," in *History of Islamic Philosophy*, ed. Seyyed Hossein Nasr and Oliver Leaman, 277–93 (London: Routledge, 1996), especially 284f, and, for a succinct explanation of its use in Jewish philosophy, see Kaufmann Kohler and Isaac Broydé, "Empedocles of Agrigentum," http://www.jewishencyclopedia.com/articles/5740-empedocles-of-agrigentum, accessed March 27, 2021.

4. Glorious Return: Resurrected Bodies

1. 1 Corinthians 15:51. Unless otherwise specified, all Biblical translations are from the New Revised Standard Version.

2. *Genesis Rabbah* 3:7, composed ca. 500 CE, Sefaria Community Translation, accessed March 27, 2021, at https://www.sefaria.org/Bereishit_Rabbah.3?lang=bi. In all notes, where the URL differs between the footnote and bibliographic references, the footnote reference is to the cited point in the text and the bibliographic reference is to the starting point of the text. Cf. *Genesis Rabbah* 8:1: "If a man is worthy enough he enjoys both worlds, for it says, 'Thou hast formed me for a later [world] and an earlier [world]'"; from *Midrash Rabbah*, ed. and trans. H. Freedman and M. Simon (London: Soncino Press, 1961). All other references to *Genesis Rabbah* are to the Sefaria translation.

3. Howard Schwartz, *A Palace of Pearls: The Stories of Rabbi Nachman of Bratslav* (Oxford: Oxford University Press, 2018), 64–65, citing twentieth-century rabbi Kalonymus Kalman Shapira. For the differences among Kabbalists regarding the creation of a series of worlds, see Aryeh Kaplan, *Immortality, Resurrection, and the Age of the Universe: A Kabbalistic View* (Brooklyn, N.Y.: KTAV, 1993), 6. The *Sefer HaTemunah* in particular is cited there as supporting the serial-worlds idea; cf. *Sanhedrin* 97a, composed ca. 450–550 CE, trans. Adin Even-Israel Steinsaltz, accessed March 27, 2021 at https://www.sefaria.org/Sanhedrin?lang=bi.

4. *Sanhedrin* 90a. Cf. *Sanhedrin* 10:1a, 11:99a. Other rabbinic arguments for resurrection occur in *Sanhedrin* 90b–92a. See also *Genesis Rabbah* 20:26; *Leviticus*

Rabbah (*Vayikra Rabbah*), trans. Mike Feuer, accessed March 27, 2021, https://www.sefaria.org/Vayikra_Rabbah?lang=bi, 27:4.

5. Flavius Josephus, *Antiquities of the Jews*, in *The Genuine Works of Flavius Josephus the Jewish Historian*, trans. William Whiston (London, 1737), 8.1.4, 17.1.3.

6. *Sanhedrin* 11:99.

7. Matthew 22:23; Mark 12:18; Luke 20:27; Acts 23:8.

8. Acts of the Apostles, 23:6, 8. Paul also identifies himself as a Pharisee in Acts 26:5.

9. Flavius Josephus, *War of the Jews*, in Whiston, *Genuine Works of Flavius Josephus the Jewish Historian*, 2.8.11. According to Josephus, the Essenes hold that all matter is impermanent, while souls are immortal; set free by death from the "bonds of the flesh," souls "rejoice and mount upward." As Josephus points out, this notion "is like the opinions of the Greeks." See, e.g., Plato's *Phaedrus*, trans. Alexander Nehamas and Paul Woodruff (Indianapolis: Hackett, 1995), where the soul is like a chariot pulled by winged horses into the orbit of the gods (246a–50c). There are of course other Greek opinions.

10. Daniel J. Silver, "The Resurrection Debate," in *Moses Maimonides' Treatise on Resurrection*, ed. Fred Rosner (Northvale, N.J.: Jason Aronson, 1997), 77.

11. *Sifrei Devarim*, compiled ca. 3rd century CE, Sefaria Community Translation, accessed March 27, 2021, at https://www.sefaria.org/Sifrei_Devarim?lang=bi, 306:35.

12. *Kiddushin*, composed ca. 45–550 CE, William Davidson Talmud, Sefaria Community Translation, accessed March 27, 2021, at https://www.sefaria.org/Kiddushin.39b.7?ven=Sefaria_Community_Translation&lang=bi. 39b:7.

13. Psalms 16:11. See also 2 Maccabees 7:9: "You dismiss us from this present life, but the King of the universe will raise us up to an everlasting renewal of life, because we have died for his laws."

14. 1 Samuel 2:6; Deuteronomy 2:6–7; "Redemption and Resurrection," 4Q521, in *The Dead Sea Scrolls: A New Translation*, trans., commentary Michael Wise, Martin Abegg Jr., and Edward Cook (New York: HarperOne, 2005), Fragments 2+4 2:1, 7–8, 12–14. Cf. Deuteronomy 32:39: "See now that I, even I, am he; there is no god besides me. I kill and I make alive; I wound and I heal; and no one can deliver from my hand."

15. Isaiah 26:19.

16. Job 19:25–26: "For I know that my Redeemer lives, and that at the last he will stand upon the earth; and after my skin has been thus destroyed, then in my flesh I shall see God."

17. Ezekiel 37:7–11. The passage echoes Job's description of his body's initial creation by God: "You clothed me with skin and flesh, and knit me together with bones and sinews"; Job 10:11. Job in turn echoes Psalm 139:13–15: "For it was you who formed my inward parts; you knit me together in my mother's womb. I praise you, for I am fearfully and wonderfully made."

18. 1 Corinthians 15:12–13.

19. 1 Corinthians 15:50–54. See also Matthew 27:52: "The tombs also were opened, and many bodies of the saints who had fallen asleep were raised"; 1 Thessalonians 4:16: "For the Lord himself, with a cry of command, with the archangel's call and

with the sound of God's trumpet, will descend from heaven, and the dead in Christ will rise first"; and Acts 26:8: "Why is it thought incredible by any of you that God raises the dead?"

20. Irenaeus, *Against Heresies*, in *Ante-Nicene Fathers*, vol. 1, ed. Alexander Roberts, James Donaldson and A. Cleveland Coxe (Buffalo, N.Y.: Christian Literature, 1886), 1.10.1.

21. This is the opening line of Tertullian's *De Resurrectione Carnis*: "Fiducia Christianorum resurrectio mortuorum: illam credentes hoc sumus. Hoc credere veritas cogit: veritatem deus aperit"; in *Patrilogia Latina* 2 (Paris: J. P. Migne, 1844). English: Alexander Roberts, James Donaldson, and A. Cleveland Coxe, eds., *Ante-Nicene Fathers*, vol. 3 (Buffalo, N.Y.: Christian Literature, 1885).

22. Tertullian, *On Resurrection*, chap. 46, with reference to 1 Corinthians 15:50.

23. *The Treatise on the Resurrection (Epistle to Rheginus)*, trans. Malcolm Peel, in *The Nag Hammadi Library*, ed. James M. Robinson (New York: HarperCollins, 1990), 44, 4–9.

24. 2 Timothy 2:18. The Valentinians, who generally accepted the same scriptural sources as any other Christians of their era, did not regard the letters to Timothy or to Titus as part of the canon; see Ismo Dunderberg, "Valentinus and His School," *Revista Catalana de Teologia* 37, no. 1 (2012): 136. Though these letters are now canonical, scholars do generally suspect that they were not written by Paul.

25. It is more likely, as he argues, that Paul's claim for the resurrection of the dead simply seemed to his educated opponents very like, and no more likely than, other stories of magicians, gods, or even physicians raising the dead; Dale Martin, *The Corinthian Body* (New Haven, Conn.: Yale University Press, 1999), 110–14.

26. Irenaeus, *Against Heresies* 2.31; Tertullian, *On Resurrection*, chaps. 15, 19, 23, 38, 49.

27. See Tertullian, *On Resurrection* chaps. 23–24; Irenaeus, *Against Heresies* 3.

28. Augustine, *City of God*, in *Nicene and Post-Nicene Fathers*, ed. Philip Schaff, series 1, vol. 2 (London: T & T Clark, 1887), 20.6; cf. *On the Trinity*, in *Nicene and Post-Nicene Fathers*, series 1, vol. 3, 4.3; Thomas Aquinas, *Summa Theologiae*, trans. Fathers of the English Dominican Province (London: Thomas Baker, 1917), Supplement, Question 77.

29. John 5:25; Augustine, *City of God* 20.6, and *Tractates on the Gospel of John*, trans. John Gibb, in *Nicene and Post-Nicene Fathers*, ed. Philip Schaff, series 1, vol. 7 (Buffalo, N.Y.: Christian Literature, 1888), 19.9.

30. Maimonides, "The 13 Principles and the Resurrection of the Dead," in *Medieval Sourcebook*, ed. Paul Halsall, revised 2019, accessed March 27, 2021, at https://sourcebooks.fordham.edu/source/rambam13.asp.

31. *Sanhedrin* 91b; cf. Irenaeus, *Against Heresies* 2.29.2.

32. "In this life men, composed of soul and body, sin or act rightly. Therefore, in both the soul and the body men deserve reward or punishment"; Thomas Aquinas, *The Summa Contra Gentiles of Thomas Aquinas*, trans. English Dominican Fathers (New York: Benzinger Brothers, 1929), 4.79.12. For the earlier Christian instances,

see Irenaeus, *Against Heresies* 2.29; Tertullian, *On Resurrection*, chap. 8. Cf. Athenagoras, *On the Resurrection*, trans. B. P. Pratten, in *Ante-Nicene Fathers*, vol. 2, ed. James Donaldson, Alexander Roberts, and A. Cleveland Coxe (Buffalo, N.Y.: Christian Literature, 1885), 18.2–5, cited in David Rankin, *The Early Church and the Afterlife: Post-Death Existence in Athenagoras, Tertullian, Origen and the Letter to Rheginos* (New York: Routledge, 2017), part 2.

33. See Maimonides, *Treatise on Resurrection*, trans. Fred Rosner (Lanham, Md.: Rowman and Littlefield, 2004).

34. Maimonides, *Mishnah Torah: Repentance* 8.2, composed ca.1176–78, trans. Simon Glazer (NP: Maimonides, 1927), accessed March 27, 2021, at https://www.sefaria.org/Mishneh_Torah,_Repentance?lang=bi.

35. See Henry Malter, *Saadia Gaon: His Life and Works* (Philadelphia: Jewish Publication Society of America, 1921), 231; Saadia Gaon, Treatise 7, "Concerning the Resurrection of the Dead in this World," in *The Book of Beliefs and Opinions*, trans. Samuel Rosenblatt (New Haven, Conn.: Yale University Press, 1976), 264–85.

36. Thomas Aquinas, *Summa Theologiae*, Supplement, Questions 69–86; *Summa Contra Gentiles*, 4.79–91.

37. Thomas Aquinas, *Summa contra Gentiles*, 4.79.11; Augustine, *City of God* 13.19.

38. John 5:24, my italics.

39. Augustine, *City of God* 13.19. Augustine makes reference here to Luke 21:18.

40. Augustine, *Confessions*, trans. Henry Chadwick (Oxford: Oxford University Press, 1991), 10.15.

41. See, for example, Augustine, *City of God* 9.17: "For by his incarnation he showed us, for our salvation, two truths of the greatest importance: that the true divine nature cannot be polluted by flesh, and that demons are not to be reckoned our superiors because they are not creatures of flesh."

42. Augustine, *City of God* 13.19.

43. Athenagoras, *On the Resurrection* 2, 14.6–15, cited and translated in Rankin, *Early Church and the Afterlife*, 92.

44. Irenaeus, *Against Heresies* 10.3.2.

45. Tertullian, *On Resurrection*, chap. 11.

46. Augustine, *City of God* 22.1.

47. *Sanhedrin* 91a.

48. *Sanhedrin* 91a.

49. Eliezer Schlossberg and Dov Schwartz note that "Sa'adia compares resurrection to creation, a comparison which is a leitmotif within several issues discussed in Treatise VII." They also point out that this comparison "is a central principle of the Muslim conception of this subject," appropriate to the context of Islamic influence in which Saadia worked; Schlossberg and Schwartz, "From Periphery to Center: Early Discussion of Resurrection in Medieval Jewish Thought," *Hebrew Union College Annual* 89 (2018): 192.

50. *Genesis Rabbah* 24; cf. *Ecclesiastes Rabbah*, composed ca. 700–950 CE, Sefaria Community Translation, accessed March 27, 2021, at https://www.sefaria.org/Kohelet_Rabbah.12.5?lang=bi, 12:5; and *Leviticus* Rabbah 18:1.

51. *Zohar* 1:137a.

52. "Rabbi Shim'on said, 'Concerning this the ancient ones disputed; but with these bones revived by the blessed Holy One, He performed miracles and remarkable signs. Come and see what is written: *Remember, now, that like clay You worked me, and to dust You will return me* (Job 10:9). What is written next? *Will You not pour me out like milk, and like cheese congeal me? With skin and flesh You will clothe me, with bones and sinews weave me* (Job 10–11). After a person has decayed on the dust—and the time of revival of the dead has arrived—the blessed Holy One will make that enduring bone like dough, flowing as *milk*; and from that pure flow, sparkling in clarity, the bone will be blended, liquefied like milk. Afterward He will congeal it, and it will be fashioned in the form of curdled *cheese*; and afterward *skin* and *flesh* and *bones* will be drawn over it"; *Zohar* 2:28b.

53. *Sifrei Devarim*, 306. Cf. 2 Maccabees 7:23: "Therefore the Creator of the world, who shaped the beginning of humankind and devised the origin of all things, will in his mercy give life and breath back to you again"; Irenaeus, *Against Heresies* 5.15.1–2; Tertullian, *Against Marcion*, bk. 2, in *Ante-Nicene Fathers*, ed. James Donaldson, Alexander Roberts, and A. Cleveland Coxe, vol. 3.

54. Isaiah 66:14. Thus cited in Tertullian, *On Resurrection*, chap. 31. The NRSV gives, "You shall see, and your heart shall rejoice; your bodies shall flourish like the grass."

55. Tertullian, *On Resurrection*, chap. 12.

56. *Zohar* 2:28b.

57. Numbers 17:8–10, New International Version. NRSV has "so that you may make an end of their complaints against me, or else they will die."

58. Nachmanides (Moses ben Nachman), *Commentary on Genesis* 1:26, composed ca. 1246–86, Sefaria Community Translation, accessed March 27, 2021, at https://www.sefaria.org/Ramban_on_Genesis?lange=bi&p2=Ramban_on_Genesis.1.26.1&lange2=en.

59. *Genesis Rabbah* 8:3.

60. *Genesis Rabbah* 8:11. The same text also suggests an alternative reading, in which humanity must be created as participating in both celestial and terrestrial, lest one element outnumber the other. Rashi finds the point again in the second creation story—"And God formed—here the letter yod is written twice to intimate that there were two formations—a formation of man for this world, and a formation of man for resurrection"; *Pentateuch with Rashi's Commentary*, trans. M. Rosenbaum and A. M. Silbermann, accessed March 27, 2021, at https://www.sefaria.org/Rashi_on_Genesis?lang=bi, on Genesis, 2:7.

61. *Sifrei Devarim* 306:15, with reference to Psalm 50:4 and Ezekiel 37.

62. 2 Baruch 50:2, trans. Robert Henry Charles, in *The Apocrypha and Pseudepigrapha of the Old Testament in English* (Oxford: Clarendon, 1913). *Genesis Rabbah* 95 suggests similarly that the dead will rise as they were at death and be healed by God afterward.

63. Genesis 3:19.

64. *Genesis Rabbah* 20:11, emphasis original. See also *Targum Pseudo-Jonathan*, on Genesis 3: "It is from the dust it is to be that thou art to arise, to render judgment and reckoning for all that thou has done, in the day of the great judgement," composed ca. 150–250 CE, from *The Targum of Jonathan ben Uzziel*, trans. J. W. Etheridge (London: Longman, Green, Longman, and Roberts, 1862), on Genesis 3.

65. *Genesis Rabbah* 21; cf. 14.

66. *The Tripartite Tractate*, trans. Einar Thomassen, in *The Nag Hammadi Scriptures*, ed. Martin Meyer (New York: HarperCollins, 2008), 117, 15–20.

67. *Gospel of Thomas*, trans. Marvin Meyer, in Meyer, *Nag Hammadi Scriptures*, log. 51.

68. *Gospel of Thomas*, log. 49.

69. Acts 3:20–21.

70. Gregory of Nyssa, *On the Soul and the Resurrection*, in *Nicene and Post-Nicene Fathers*, series 2, vol. 5 (Grand Rapids, Mich.: William B. Eerdmans, 1888), 797. Cf. 839: "In any and every case evil must be removed out of existence, so that . . . the absolutely non-existent should cease to be at all. Since it is not in its nature that evil should exist outside the will, does it not follow that when it shall be that every will rests in God, evil will be reduced to complete annihilation, owing to no receptacle being left for it?"

71. Jerome, *Commentary on Ephesians* 4:4, in *History of Opinions on the Scriptural Doctrine of Retribution*, ed. Edward Beecher (New York: D. Appleton, 1878), 263. Oddly, Jerome holds this view despite his dislike of the way that some versions of apokatastasis undo hierarchies, even those of virtue. See the discussion in Jennifer Glancy, *Corporal Knowledge: Early Christian Bodies* (Oxford: Oxford University Press, 2010), 70.

72. *Gospel of Philip* 67, 10–20.

73. Origen, *On First Principles*, trans. G. W. Butterworth (Gloucester, Mass.: Peter Smith, 1973), 1.2.2: "And because in this very subsistence of wisdom there was implicit every capacity and form of the creation that was to be, both of those things that exist in a primary sense and of those which happen in consequence of them, the whole being fashioned and arranged beforehand by the power of foreknowledge, wisdom, . . . in regard to these very created things that had been as it were outlined and prefigured in herself, says that she was created as a 'beginning of the ways' of God, which means that she contains within herself both the beginnings and the causes and species of the whole creation." See also 1.6.2: "For the end is always like the beginning; as therefore there is one end of all things, so we must understand that there is one beginning of all things, and as there is one end of many things, so from one beginning arise many differences and varieties, which in their turn are restored." Alexander Altmann also notes that for Ibn Ezra, emanation is the "innovation" into time and into matter (which already exists) of the essences that already exist in the Divine Mind; Altmann, "A Note on the Rabbinic Doctrine of Creation," *Journal of Jewish Studies* 7, no. 3–4 (1956): 204.

74. Origen, *Commentary on the Gospel According to John*, books 1–10, trans. Ronald E. Heine (Washington, D.C.: The Catholic University of America Press, 1989), chap. 39.

75. 1 Cor 15:25–28.

76. Origen, *On First Principles* 1.6.4: "For if the heavens shall be 'changed,' certainly that which is 'changed' does not perish; and if 'the form of this world passes away,' it is not by any means an annihilation or destruction of the material substance that is indicated, but the occurrence of a certain change of quality and an alteration of the outward form."

77. See, for many examples, *Mishnah Gittin*, composed in Talmudic Israel (c. 190–c. 230 CE), trans. Rabbi Shraga Silverstein, accessed January 2, 2020, at https://www.sefaria.org/Mishnah_Gittin?lang=bi. Also see, among other texts, *Mishnah Ketubot*, composed ca.190–230 CE, trans. Aden Even-Israel Steinsaltz, accessed March 72, 2021, at https://www.sefaria.org/Mishnah_Ketubot?lang=bi, chap. 12.

78. *The Beginning of Wisdom*, trans. Amiram Markel and Michael Tzvi Wolkenfeld, 2004–2006. Accessed March 27, 2021, at https://www.sefaria.org/The_Beginning_of_Wisdom?lang=bi: "Within the framework of time, there are six basic stages or branches. The first stage was the creation of the world through the name of 52-Ban, representing strict judgment. This was also the world of *Tohu*—Chaos or *Nekudim*—Points. The second stage was the introduction of the name 45-Mah representing mercy. This is the world of *Tikkun*—Repair"; chap. 8, §159. Shortly before Nachmanides's work, Isaac the Blind describes an extensive series of ascents and reembodiments of the soul, aimed toward an ultimate judgement and resurrection in the body.

79. *Zohar* 3:31b.

80. Though I do not explore it here, it may be worth noting that a further option is the multiplicity of the soul. In his *Address to the Greeks*, the second-century Christian Tatian writes, "The human soul consists of many parts, and is not simple; it is composite, so as to manifest itself through the body; for neither could it ever appear by itself without the body, nor does the flesh rise again without the soul"; Tatian, "Address to the Greeks," trans. J. E. Ryland, in *Ante-Nicene Fathers*, vol. 2, chap. 15.

81. For instance, two of the translators of the *Epistle*, Malcolm Peel and Bentley Layton, respectively argue that it presents a bodily resurrection and one of soul alone; Mark J. Edwards, "The Epistle to Rheginus: Valentinianism in the Fourth Century," *Novum Testamentum* 3, no. 1 (January 1995): 87.

82. *Gospel of Philip*, trans. Marvin Meyer, in Marvin Meyer, ed., *Nag Hammadi Scriptures* (New York: HarperCollins, 2008), 157–86, 57, 10.

83. *Gospel of Philip* 57, 11–20.

84. *Gospel of Philip* 56, 25–35, with reference to 1 Cor 15:50.

85. Jorunn Jacobsen Buckley, "A Cult-Mystery in *The Gospel of Philip*," *Journal of Biblical Literature* 99, no. 4 (1980): 570.

86. *Treatise on the Resurrection* 48, ca. 10–15.

87. Buckley, "Cult-Mystery," 151; citations from *Gospel of Philip* 56 and 57.

88. Section II of T. S. Eliot's "Ash Wednesday," in which this line occurs ("And neither division nor unity/Matters"), also draws extensively on imagery from Ezekiel's story of the risen bones; "Ash Wednesday," in *T. S. Eliot: Collected Poems, 1909–1962* (New York: Harcourt, Brace, 1991), 85–96.

89. *Gospel of Philip* 74, 12–20.

90. *Gospel of Philip* 67, 25–27.

91. *Gospel of Philip* 67, 27–30; cf. *Gospel of Thomas* log. 22, where Jesus tells his disciples, "When you make the two into one, and when you make the inner like the outer and the outer like the inner and the upper like the lower, and when you make male and female into a single one, so that the male will not be male nor the female be female, . . . then you will enter the kingdom."

92. *Gospel of Philip* 85, 15–19.

93. "The master [did] everything in a mystery: baptism, chrism, eucharist, redemption, and bridal chamber"; *Gospel of Philip* 67, 27–30.

94. *Treatise on the Resurrection* 49, 20–25.

95. *Treatise on the Resurrection* 47, 25–30.

96. *Gospel of Philip* 68, 29–37.

97. *Treatise on the Resurrection* 49, 5–10.

98. M. J. Edwards, "Epistle to Rheginus," 82–83. Also see Elaine Pagels, *The Gnostic Paul: Gnostic Exegesis of the Pauline Letters* (Philadelphia: Fortress, 1975).

99. *Treatise on the Resurrection* 48, 30–49, 5.

100. Origen, *Against Celsus*, trans. Frederick Cromble, in *Ante-Nicene Fathers*, vol. 4, 7.32.

101. Origen, *Against Celsus* 20, 21; Porphyry, "The Life of Pythagoras," trans., ed. Kenneth S. Guthrie and David Fideler, in *The Pythagorean Sourcebook and Library: An Anthology of Ancient Writings Which Relate to Pythagoras and Pythagorean Philosophy* (Grand Rapids, Mich.: Phanes, 1987), par. 19.

102. Origen, *Against Celsus* 7.32.

103. In 1 Cor 15:35–38 and again at 15:42–44, Paul uses agricultural imagery, declaring that the body is sown perishable or corruptible, and rises imperishable or incorruptible.

104. Origen, *Against Celsus* 7.32.

105. "Not all flesh is alike, but there is one flesh for human beings, another for animals, another for birds, and another for fish. There are both heavenly bodies and earthly bodies, but the glory of the heavenly is one thing, and that of the earthly is another. There is one glory of the sun, and another glory of the moon, and another glory of the stars; indeed, star differs from star in glory. So it is with the resurrection of the dead"; 1 Cor 15:39–42. In *The Corinthian Body*, Dale Martin cites Alan Scott, who writes that the idea of mind/soul as a celestial substance was "part of Hellenistic folklore"; see Alan Scott, *Origen and the Life of the Stars: A History of an Idea* (Oxford: Clarendon, 1991).

106. See Origen, *On Prayer*, trans. William A. Curtis (Grand Rapids, Mich.: Christian Classics Ethereal Library, 2001), 31.3. Here Origen is close to the descriptions of the soul's embodiment in the neo-Platonic theurgical texts of

Iamblichus, for whom we lose our original, perfectly spherical forms when our souls and bodies become mortal and regain both roundness and immortality together. See Gregory Shaw, "Theurgy and the Platonist's Luminous Body," in *Practicing Gnosis: Ritual, Magic, Theurgy, and Liturgy in Nag Hammadi, Manichaean and Other Ancient Literature; Essays in Honor of Birger A. Pearson*, ed. April DeConick, Gregory Shaw, and John D. Turner (Leiden: Brill, 2013), 545–47.

107. Though Origen resists the identically returning cosmoses of Pythagoreans and Stoics, he does declare that "surely the end of this world is the beginning of the one to come," which could argue for another rebirth. And, as Virginia Burrus points out, he even tells multiple stories of different beginnings. Both points are in Burrus, *Ancient Christian Ecopoetics: Cosmologies, Saints, Things* (Philadelphia: University of Pennsylvania Press, 2018), 38, citing Origen, *On First Principles* 2.1.3.

108. A useful summary can be found in Daniel J. Silver, "Nachmanides' Commentary on the Book of Job," *Jewish Quarterly Review* 60, no. 1 (1969): 22–25.

109. "The rationale behind reincarnation or transmigration is dealt with in the Zohar in a long passage called *Saba d'Mishpatim*. . . . The central idea is that reincarnation, or *gilgul*, has two purposes: (a) to rectify sin; (b) to acquire higher levels of soul"; "The Secret of Servitude," part 1, From the teachings of Rabbi Shimon bar Yochai; translation and commentary by Moshe Miller, accessed March 27, 2021, at https://www.chabad.org/kabbalah/article_cdo/aid/380542/jewish/The-Secret-of-Servitude.htm.

110. Thus cited and translated in Elliot R. Wolfson, *Language, Eros, Being: Kabbalistic Hermeneutics and Poetic Imagination* (New York: Fordham University Press, 2005), 253, from Nachmanides, *Sha'ar ha-Gemul* 2:304–5.

111. Wolfson, *Language, Eros, Being*, 253n424.

112. Wolfson, *Language, Eros, Being*, 253n424.

113. *Zohar* 1:186b, in *Zohar*, vol. 3, trans. with commentary by Daniel C. Matt (Stanford, Calif.: Stanford University Press, 2006). On reincarnation, Matt cites as well *Bahir* 135 (195), *Zohar Chadash* 89b, 59a–c, and *Zohar* 1:131a, 2:94b–114a passim.

114. See Isaac Luria, *Gate of Reincarnations*, recorded by Chaim Vital, trans. Yitzchok bar Chaim, commentary by Shabtai Teicher, accessed March 27, 2021, at https://www.chabad.org/kabbalah/article_cdo/aid/378771/jewish/Gate-of-Reincarnations.htm, 11:2–3.

115. Genesis 5:1–2.

116. *Midrash Tanhuma-Yelammedenu: An English Translation of Genesis and Exodus from the Printed Version of Tanhuma-Yelammedenu with an Introduction, Notes, and Indexes*, trans. Samuel A. Berman (Hoboken, N.J.: KTAV, 1996), glossing Job 38:4, accessed January 3, 2020, at https://www.sefaria.org/Midrash_Tanchuma%2C_Ki_Tisa.12.1?lang=bi&with=all&lang2=en.

117. Luria, *Gate of Reincarnations* 3.2, accessed March 27, 2021.

118. Michael Laitman, *Unlocking the Zohar* (Toronto: Laitman Kabbalah, 2011), 57: "Before the Creator created this world, He created worlds and destroyed them. This is the breaking of the vessels. Finally, the Creator desired to create this world,

and consulted with the Torah, the middle line. Then He was corrected in His corrections, decorated in His decorations, and created this world. And then, everything that exists in this world was before Him at the time of creation, and was established before Him." The relevant scriptural text is Ecclesiastes 3:15, "That which is has been already, and that which will be has already been."

119. *Zohar* 2:26b.

120. Luria, *Gate of Reincarnations* 1:1.

121. Lawrence Fine, "Tikkun: A Lurianic Motif in Contemporary Jewish Thought," in *From Ancient Israel to Modern Judaism: Essays in Honor of Marvin Fox*, ed. Jacob Neusner, Ernest S. Frerichs, and Nahum M. Sarna (Providence, R.I.: Brown University Press, 1989), 35–54.

122. See Luria, *Gate of Reincarnations*, esp. 11:7.

123. See Matthew 22:23–33; Mark 12:18–27; Luke 20:27–38.

124. Luria, *Gate of Reincarnations* 5:5.

125. I do not take up here the complications introduced by the concept of multiple souls in one body, which can occur at birth or when a person enters adulthood. In both cases, the aim is still to perfect the soul(s) in embodied action; see Luria, *Gate of Reincarnations* 5:1, 5:2.

126. See useful summaries of this point at Yimiyahu Ullman, "Reincarnation," accessed March 27, 2021, at https://ohr.edu/1077, and Shimon bar Yochai, "Resurrection and Reincarnation," trans. and commentary by Shabtai Teicher, accessed March 27, 2021, at https://www.chabad.org/kabbalah/article_cdo/aid/380668/jewish/Resurrection-and-Reincarnation.htm.

127. Athenagoras, *On the Resurrection*, chap. 2.

128. Athenagoras, *On the Resurrection*, my emphasis.

129. Athenagoras, *On the Resurrection*, chap. 25.

130. Gregory of Nyssa, *On the Soul*, 844.

131. Constitutions of the Fourth Lateran Council, chap. 1, "Confession of Faith," Accessed March 27, 2021, at https://www.papalencyclicals.net/councils/ecum12-2.htm#1.

132. Aquinas offers more extensive argument for the natural unity of body and soul in *Summa contra Gentiles* 4.79.10.

133. Thomas Aquinas, *Summa Theologiae*, Supplementum, q. 75, art. 3: Whether the resurrection will be for all without exception. Athenagoras makes a very similar argument in *On the Resurrection*, chap. 15.2–7.

134. See Thomas Aquinas, *Summa Theologiae*, Supplementum, q 75, art. 1, which argues that Aristotle has demonstrated that soul and body are joined as form and matter. "Hence it is clear that if *man* cannot be *happy* in this life, we must of *necessity* hold the *resurrection*."

135. Thomas Aquinas, *Summa contra Gentiles* 4.84.7: "If the body of the man who rises is not to be composed of the flesh and bones which now compose it, the man who rises will not be numerically the same man."

136. Thomas Aquinas, *Summa contra Gentiles* 4.82.9.

137. Augustine, *City of God* 22.12–13.

138. For all of this discussion, see Augustine, *City of God* 22.14–15.

139. Augustine, *City of God* 22.14: "For in the Lord's words, where He says, Not a hair of your head shall perish, it is asserted that nothing which was possessed shall be wanting; but it is not said that nothing which was not possessed shall be given. To the dead infant there was wanting the perfect stature of its body; for even the perfect infant lacks the perfection of bodily size, being capable of further growth. This perfect stature is, in a sense, so possessed by all that they are conceived and born with it—that is, they have it potentially, though not yet in actual bulk; just as all the members of the body are potentially in the seed, though, even after the child is born, some of them, the teeth for example, may be wanting."

140. Thomas Aquinas, *Summa contra Gentiles* 4.81.12.

141. Augustine, *City of God* 22.19.

142. Augustine, *City of God* 22.20.

143. Schlossberg and Schwartz, "From Periphery to Center," 191–92. Reference to Saadia Gaon, רס״ג ופירוש עם תרגום אנטיוכס ומגילת דניאל וספר, ed. Y. Qafih (Jerusalem: Committee for the Publication of the books of R. Sa'adia Gaon 1981), 227.

144. In Schlossberg and Schwartz, "From Periphery to Center," 189. Reference to Saadia Gaon, in L. Gardet, "Ḳiyāma," in *Encyclopaedia of Islam*, 2nd ed., ed. P. Bearman, Th. Bianquis, C. E. Bosworth, E. van Donzel, and W. P. Heinrichs, accessed March 27, 2021, at https://referenceworks.brillonline.com/browse/encyclopaedia-of-islam-2, (a) III.

145. Augustine, *City of God* 22.20.

146. Tertullian writes in relation to the story of Jonah, "You will ask, Will then the fishes and other animals and carnivorous birds be raised again, in order that they may vomit up what they have consumed, on the ground of your reading in the law of Moses, that blood is required of even all the beasts? Certainly not. But the beasts and the fishes are mentioned in relation to the restoration of flesh and blood, in order the more emphatically to express the resurrection of such bodies as have even been devoured"; *On Resurrection* 32. Athenagoras, *On the Resurrection* 8.

147. Athenagoras, *On the Resurrection* 4.

148. Athenagoras, *On the Resurrection* 5–8.

149. Augustine, *City of God* 22.20.

150. Thomas Aquinas, *Summa contra Gentiles* 4.81.13.

151. Augustine, *City of God* 22.20.

152. Thomas Aquinas, *Summa Theologiae*, Supplementum, 80.1.

153. Thomas Aquinas, *Summa Theologiae*, Supplementum, 80.3.

154. Unexpectedly, I was reminded of this path of thought by a Twitter thread, presented by medievalist Erik Wade in July 2019, thus demonstrating the value of procrastination; accessed March 27, 2021, at https://twitter.com/erik_kaars/status/1139106297339686912.

155. Thomas Aquinas, *Summa contra Gentiles* 4.80.5.

156. Thomas Aquinas, *Summa contra Gentiles* 4.81.13.

157. Thomas Aquinas, *Summa contra Gentiles* 4.81.13.

158. Origen, *Against Celsus* 5.21. Friedrich Nietzsche notes the association with Pythagoreanism: "At bottom, indeed, that which was once possible could present itself as a possibility for the second time only if the Pythagoreans were right in believing that when the constellation of the heavenly bodies repeated the same things, down to the smallest event, must also be repeated on earth"; Nietzsche, *Untimely Meditations*, trans. R. J. Hollingdale (Cambridge: Cambridge University Press, 1997), 69–70. Like Origen, Porphyry attributes the idea of exact recurrence to Pythagoras, alongside the idea of recurrent incarnations; Porphyry, "Life of Pythagoras," chap. 19.

159. In his unfinished *Philosophy in the Tragic Age of the Greeks*, Nietzsche also finds the concept of cosmic recurrence in both Heraclitus (with whom he continues to feel some affinity) and the Milesian Anaximander, both of whom, he says, believe "in a periodically repeated end of the world, and in an ever renewed rise of another world out of the all-destroying cosmic fire"; Nietzsche, *Philosophy in the Tragic Age of the Greeks*, trans. Marianne Cowan (Washington, D.C.: Regnery, 1962), 60. Nietzsche attributes the Stoic version ultimately to Heraclitus; see Nietzsche, *Ecce Homo: How One Becomes What One Is*, trans. Walter Kaufmann and R. J. Hollingdale (New York: Penguin Classics, 1992), "The Birth of Tragedy," 3.

160. Nemesius of Emesa, *On the Nature of Man* 309.5–311.2, in *The Hellenistic Philosophers*, vol. 1: *Translations of the Principal Sources, with Philosophical Commentary*, ed., trans. A. A. Long and D. N. Sedley (Cambridge: Cambridge University Press, 1987), entry 52 C.

161. Nietzsche, *The Gay Science*, trans. Walter Kaufmann (New York: Vintage, 1974), §341.

162. Empedocles, frag. B17, in *A Presocratics Reader: Selected Fragments and Testimonia*, ed. Patricia Curd (Indianapolis: Hackett, 2011). Nietzsche writes of his admiration for Empedocles, among others, in *Philosophy in the Tragic Age of the Greeks*, 31. Of the Nietzschean work most concerned with eternal recurrence, Babette Babich writes, "It has traditionally been observed that the figure of Empedocles is key to Nietzsche's *Zarathustra*"; Babette Babich, "The Philosopher and the Volcano: On the Antique Sources of Nietzsche's Übermensch," *Philosophy Today* 55, Supplement (2011): 206.

163. Nietzsche, *The Will to Power*, trans. Walter Kaufmann and R. J. Hollingdale (New York: Vintage, 1968), §1067: "Supposing that the world has a certain quanta of force at its disposal, then it is obvious that every displacement of power at any point would affect the whole system—thus together with the sequential causality there would be a contiguous and concurrent dependence."

164. Nietzsche, *Ecce Homo*, "Thus Spoke Zarathustra," 1.

165. In the letter Nietzsche declares, "Thoughts have arisen such as I have never seen before," such that "the intensities of my feeling make me shudder and laugh," leaving his eyes inflamed by "tears of joy"; Nietzsche, letter to Peter Gast, August 14, 1881; Letter 90 in *The Selected Letters of Friedrich Nietzsche*, ed., trans. Christopher Middleton (Indianapolis: Hackett, 1969).

166. Pierre Klossowski, "Nietzsche's Experience of the Eternal Return," in *The New Nietzsche*, ed. David B. Allison (New York: Dell, 1977), 119.

167. Klossowski, "Nietzsche's Experience," 107–8.

168. See Nietzsche, *Gay Science*, §341. In his autobiographical work *Ecce Homo*, Nietzsche writes, "But I confess that the deepest objection to the 'Eternal Recurrence,' my real idea from the abyss, is always my mother and my sister." This suggests that Nietzsche regarded the affirmation of recurrence as an affirmation of a whole lifetime, and that there were some elements of his life that made it hard for him to affirm the sudden thought of recurrence that was brought about by beauty; *Ecce Homo*, "Why I Am So Wise," 3.

169. Klossowski, "Nietzsche's Experience," 108.

170. Klossowski, "Nietzsche's Experience," 108–9: "I de-actualize my present self to will myself in all the other selves, whose entire series must be gone through so that following the circular movement I can again become what I am at the moment in which I discover the law of the Eternal Return." He adds later, "Subsequent reflection declares: If this thought gained control over you it would make of you an other" (115).

171. Gregory of Nyssa, *On the Soul*.

172. Origen, *On First Principles* 2.2.2.

173. Origen, *Commentary on Matthew*, as cited and translated in Carolyn Walker Bynum, *The Resurrection of the Body in Western Christianity, 200–1336* (New York: Columbia University Press, 1995), 67n28, from E. Benz and E. Klostermann, eds., *Origenes Werke*, vol. 10 (Leipzig: Hinrichs, 1935), book 17, chaps. 29–30, pp. 667–71.

174. Malter, *Saadia Gaon*, 225.

175. Thomas Aquinas, *Summa contra Gentiles* 4.86–89.

176. *Midrash Tanhuma*, on Genesis 6:1.

177. See, e.g., Daniel 12:2 ("The enlightened will blaze like the brilliance of the firmament, and they who make the many righteous like the stars for ever and ever"); Judges 5:31 ("They that love Him shall be as the sun that goeth forth in its might"); 1 Enoch 38:3 ("The Lord of Spirits has caused His light to appear on the face of the holy, righteous, and elect," such that "before Him [they] shall be strong as fiery lights"); 2 Enoch 22:8–9 ("He anointed me, and dressed me, and the appearance of that ointment is more than the great light, . . . shining like the sun's ray, and I looked at myself, and was like one of his glorious ones"); 1 Enoch and 2 Enoch, in Charles, *The Apocrypha and Pseudepigrapha of the Old Testament in English*.

178. Gregory of Nyssa, *On the Soul*.

179. 2 Baruch 51:2–3: "For the aspect of those who now act wickedly shall become worse than it is, as they shall suffer torment. Also (as for) the glory of those who have now been justified in My law, who have had understanding in their life, and who have planted in their heart the root of wisdom, then their splendor shall be glorified in changes, and the form of their face shall be turned into the light of their beauty."

180. Origen, *Commentary on Romans* 5.8.13, cited and translated in Thomas Dwight McGlothlin, "Raised to Newness of Life: Resurrection and Moral

Transformation in Second- and Third-Century Christian Theology" (PhD diss., Duke University, 2015), 258, from Caroline P. Hammond Bammel, "Die fehlehnden Bände des Römerbriefkommentars des Origenes," in Lothar Lies, *Origeniana Quarta* (Innsbruck: Tyrolia, 1987), 16–20.

181. *Gospel of Philip* 60, 20–61, 35.
182. Augustine, *City of God* 22.19.
183. Augustine, *City of God* 22.17. Jerome had already remarked wearily on these and related queries: "And to those of us who ask whether the resurrection will exhibit from its former condition hair and teeth, the chest and the stomach, hands and feet, and other joints, then, no longer able to contain themselves and their jollity, they burst out laughing and adding insult to injury they ask if we shall need barbers, and cakes, and doctors, and cobblers, and whether we believe that the genitalia of which sex would rise, whether our (men's) cheeks would rise rough, while women's would be soft and whether the bodies would be differentiated based on sex. Because, if we surrender this point, they immediately proceed to female genitalia and everything else in and around the womb. They deny that singular members of the body rise, but the body, which is constituted from members, they say rises"; *Letters* 84.5, in J. N. D. Kelly, *Jerome: His Life, Writings, and Controversies* (Peabody, Mass.: Hendrickson, 1998), 198.
184. Augustine, *City of God* 22.17.
185. 1 Enoch 30:7.

Afterword

1. See René Descartes, *Meditations on First Philosophy*, trans. Donald Cress (Indianapolis: Hackett, 1993), Meditation 4, "Concerning the True and the False," esp. 40.

2. As Elizabeth Agnew Cochran notes, Stoicism also entails moral assent, in which the assent responds to the goodness of divine presence in the world, moving Augustine still closer to a Stoic understanding, though his will always be a more passionate version; Cochran, "Virtuous Assent and Christian Faith: Retrieving Stoic Virtue Theory for Christian Ethics," *Journal of the Society of Christian Ethics* 30, no.1 (Spring–Summer 2010): 120. In any case, the relation of impulse or desire to assent in Stoicism is complex, and no single description can be drawn from the range of Stoic writings.

3. Peter Brown, *Augustine of Hippo: A Biography* (Berkeley: University of California Press, 1967), 149, 148. Delight in God is also central to Augustine's sense of faith; his famous conversion to Christianity is a turn not in his intellectual convictions but in his desire (*Confessions* 7.8.19–7.12.30). See also Phillip Cary, *Inner Grace: Augustine in the Traditions of Plato and Paul* (Oxford: Oxford University Press, 2008), 15–16. Cary cites Brown, *Augustine*, 148–49, and points out that "Delight was central to Augustine's understanding of love and beauty as early as *De Musica* 6:29–33."

4. *Gospel of Truth*, trans. Marvin Meyer, in Marvin Meyer, ed., *Nag Hammadi Scriptures* (New York: HarperCollins, 2008), 16, 1.
5. *Gospel of Philip*, trans. Marvin Meyer, in Meyer, *Nag Hammadi Scriptures*, 77, 15–30.
6. *Gospel of Philip* 77, 15–20.
7. *Gospel of Philip* 54, 15–19.
8. *Gospel of Philip* 77, 25–30.
9. *Gospel of Philip* 77, 30–35.
10. *Gospel of Philip* 61, 35–62, 1: "Faith receives, love gives. No one can receive without faith, and no one can give without love."
11. *Gospel of Philip* 77, 35–78, 5. The gospel goes further: "If the people who are anointed leave them and go away, the others who are not anointed but are only standing around are stuck with their own bad odor"; 78, 5–10.
12. See Plato, *Symposium*, trans. Alexander Nehemas and Paul Woodruff (Indianapolis: Hackett, 1989), 201a–12c; *Phaedrus*, trans. Alexander Nehamas and Paul Woodruff (Indianapolis: Hackett, 1995), 255c–56d.
13. Aristotle, *Metaphysics*, trans. Hugh Tredennick (London: William Heinemann, 1989), 12.1072a–73a.
14. Margaret Miles, *Desire and Delight: A New Reading of Augustine's Confessions* (New York: Crossroad, 1992).
15. For love as lack in Plato, see *Symposium* 199d–200e. This idea is implicit in Freud's understanding of the aims and objects of unconscious drives. It is developed more explicitly, and in a more obviously philosophical sense, by Jacques Lacan, who calls desire "a relation of being to lack." In Freud, see especially *Three Essays on the Theory of Sexuality* (1905) in the Standard Edition, trans. James Strachey (London: Hogarth, 1953), 7:124–245, and *Instincts and their Vicissitudes* (1915), in the Standard Edition, trans. James Strachey (London: Hogarth, 1957), 14:111–40. In Lacan, see *The Seminars*, Book II, *The Ego in Freud's Theory and in the Techniques of Psychoanalysis, 1954–55*, trans. Sylvana Tomaselli (New York: W. W. Norton, 1988), 223.
16. I read Hegel here through the interpretation of Alexandre Kojève, as did both Merleau-Ponty and Lacan. In his *Introduction to the Reading of Hegel*, Kojève summarizes: "There must be in Man ... *negating* Desire, and hence *Action* that *transforms* the given being. The human I must be an I of *Desire*—that is, an *active* I, a *negating* I, an I that *transforms* Being and creates a new being by destroying the given being"; Kojève, *Introduction to the Reading of Hegel: Lectures on the Phenomenology of Spirit*, trans. James H. Nichols Jr. (Ithaca N.Y.: Cornell University Press, 1980), 38.
17. Robert Pogue Harrison, *Gardens: An Essay on the Human Condition* (Chicago: University of Chicago Press, 2008), 12.
18. Frank O'Hara, "Meditations in an Emergency," in *Meditations in an Emergency* (New York: Grove, 1957), 38.

Bibliography

Rabbinic Sources: Midrash, Talmud, Commentary

Bava Batra. Translated by Adin Even-Israel Steinsaltz. Accessed March 27, 2021. https://www.sefaria.org/Bava_Batra?lang=bi.

The Beginning of Wisdom. Translated by Amiram Markel and Michael Tzvi Wolkenfeld. 2004–2006. Accessed March 27, 2021. https://www.sefaria.org/The_Beginning_of_Wisdom?lang=bi.

Berakhot. Translated by Adin Even-Israel Steinsaltz. Accessed February 27, 2020. https://www.sefaria.org/Berakhot?lang=bi.

Chronicles of Jerahmeel. Translated by Moses Gaster. London: Oriental Translation Fund, 1899.

Derech Eretz-Zutta. In *New Edition of the Babylonian Talmud*, edited by Isaac Mayer and Godfrey Taubenhaus Wise. Boston: Talmud Society, 1918.

Ecclesiastes Rabbah (Kohelet Rabbah). Sefaria Community Translation. Accessed March 27, 2021, at https://www.sefaria.org/Kohelet_Rabbah?lang=bi.

The Fathers According to Rabbi Nathan. Translated by Jacob Neusner. Atlanta: Scholars, 1986.

Genesis Rabbah. Translated by H. Freedman. London: Soncino, 1961.

Genesis Rabbah (Bereishit Rabbah). Sefaria Community Translation. Accessed March 27, 2021, https://www.sefaria.org/Bereishit_Rabbah?lang=bi.

Ketubot. Translated by Adin Even-Israel Steinsaltz. Accessed February 27, 2020, https://www.sefaria.org/Ketubot?lang=bi.

Kiddushin. Sefaria Community Translation. Accessed March 27, 2021, https://www.sefaria.org/Kiddushin?lang=bi.

Leviticus Rabbah (Vayikra Rabbah). Translated by Mike Feuer. Accessed March 27, 2021, https://www.sefaria.org/Vayikra_Rabbah?lang=bi.

Midrash Tanhuma-Yelammedenu. Translated by Samuel A. Berman. Hoboken, N.J.: KTAV, 1996.

Mishnah Gittin. Translated by Shraga Silverstein. Accessed March 27, 2021, https://www.sefaria.org/Mishnah_Gittin?lang=bi.

Sanhedrin. Translated by Adin Even-Israel Steinsaltz. Accessed March 27, 2021, https://www.sefaria.org/Sanhedrin?lang=bi.

Sifrei Devarim. Sefaria Community Translation. Accessed March 27, 2021, https://www.sefaria.org/Sifrei_Devarim?lang=bi.

The Targum Pseudo-Jonathan, as The Targum of Jonathan Ben Uzziel. Translated by J. W. Etheridge. London: Longman, Green, Longman, and Roberts, 1862.

Kabbalah

Sefer ha-temunah. Lemberg, 1892.

Wisdom of the Kabbalah. Translated and with commentary by S. L. Macgregor Mathers. New York: Citadel Press, 2001.

Zohar. Translated and with commentary by Daniel C. Matt. Stanford, Calif.: Stanford University Press, 2003.

Nag Hammadi Sources

The Apocryphon of John. Translated by Frederick Wisse. In *Nag Hammadi Library*, edited by James Robinson, 104–23. New York: HarperCollins, 1990. Also as: *The Secret Book of John: The Gnostic Gospel Annotated and Explained.* Translated by Stevan Davies. Woodstock, Vt.: SkyLight Paths, 2005; and "The Secret Book of John." Translated by Marvin Meyer. In *Nag Hammadi Scriptures*, edited by Marvin Meyer, 103–32. New York: HarperCollins, 2008.

The Book of Thomas. Translated by Marvin Meyer. In Meyer, *Nag Hammadi Scriptures*, 235–46.

The Gospel of Philip. Translated by Marvin Meyer. In Meyer, *Nag Hammadi Scriptures*, 157–86.

The Gospel of Thomas. Translated by Marvin Meyer, In Meyer, *Nag Hammadi Scriptures*, 139–53.

The Gospel of Truth. Translated by Marvin Meyer. In Meyer, *Nag Hammadi Scriptures*, 31–47.

The Letter to Rheginus, as The Treatise on the Resurrection (Epistle to Rheginus). Translated by Malcolm Peel. In Robinson, *Nag Hammadi Library*, 52–57.

On the Origin of the World. Translated by Marvin Meyer, In Meyer, *Nag Hammadi Scriptures*, 199–222.

The Tripartite Tractate. Translated by Harold W. Attridge and Dieter Muller. In Robinson, *Nag Hammadi Library*, 58–103.

A Valentinian Exposition, With Valentinian Liturgical Writings. Translated by Einar Thomassen and Marvin Meyer. In Meyer, *Nag Hammadi Scriptures*, 663–78.

Zostrianos. Translated by John D. Turner. In Meyer, *Nag Hammadi Scriptures*, 537–83.

Pseudepigraphic and Apocryphal Sources

1 Enoch, as The Book of Enoch. Translated by Robert Henry Charles. In *The Apocrypha and Pseudepigrapha of the Old Testament in English*, edited by Robert Henry Charles, 163–281. Oxford: Clarendon Press, 1913.
2 Baruch, as The Book of the Apocalypse of Baruch the Son of Neriah. Translated by Robert Henry Charles. In Charles, *The Apocrypha and Pseudepigrapha of the Old Testament in English*, 470–526.
2 Enoch, as The Book of the Secrets of Enoch. In *The Forgotten Books of Eden*, edited by Rutherford H. Platt, 81–106. New York: Alpha House, 1926.
The Apocalypse of Adam. Translated by G. MacRae. In *The Old Testament Pseudepigrapha*, edited by James H. Charlesworth, 1:707–20. Peabody, Mass.: Hendrickson, 1983.
The Books of Jeu and the Untitled Text in Bruce Codex. Edited by Carl Schmidt. Translated by Violet McDermot. Leiden: Brill, 1978.
"Redemption and Resurrection," 4Q521. Translated by Martin Abegg Jr., Michael Wise, and Edward Cook. In *The Dead Sea Scrolls*, 530–31. New York: HarperCollins, 2005.
The Sibylline Oracles. Translated by Milton Spenser Terry. West Bloomfield, Mich.: Franklin Classics, 2018.

Conciliar Document

Constitutions of the Fourth Lateran Council. Accessed March 27, 2021, at https://www.papalencyclicals.net/councils/ecum12-2.htm#1.

Attributed Sources

Adorno, Theodor, and Max Horkheimer. *Dialectic of the Enlightenment: Philosophical Fragments*. Translated by Edmund Jephcott. Stanford, Calif.: Stanford University Press, 2002.
Agamben, Giorgio. *The Open: Man and Animal*. Translated by Kevin Attell. Stanford, Calif.: Stanford University Press, 2003.
Altmann, Alexander. "Gnostic Themes in Rabbinic Cosmology." In *Essays in Honour of the Very Rev. Dr. J.h. Hertz*, edited by E. Levine, I. Epstein, and C. Roth, 28–52. London: E. Goldston, 1942.
———. "A Note on the Rabbinic Doctrine of Creation," *Journal of Jewish Studies* 7, no. 3–4 (1956): 204.
———. *Studies in Religious Philosophy and Mysticism*. Ithaca, N.Y.: Cornell University Press, 1969.
Andruska, Jennifer L. *Wise and Foolish Love in the Song of Songs*. Leiden: Brill, 2018.
Annas, Julia. *Aristotle's Metaphysics Books M and N*. Oxford: Oxford University Press, 1976.
Aristotle. *De Anima (on the Soul)*. Translated by R. D. Hicks. Cambridge: Cambridge University Press, 1907.

———. *Metaphysics*. Translated by Hugh Tredennick. London: William Heinemann, 1989.
———. *On the Heavens*. Translated by W. K. C. Guthrie. Cambridge, Mass.: Harvard University Press, 1939.
———. *Physics*. Translated by Robin Waterfield. Oxford: Oxford University Press, 2008.
ben Asher, Rabbi Jacob. *Tur HaAroch*. Translated by Eliyahu Munk. Jerusalem: Urim, 2005.
Athanasius. *The Life of Antony and the Letter to Marcellinus*. Translated by Robert C. Gregg. Mahwah, N.J.: Paulist Press, 1980.
Athenagoras. *A Plea for the Christians*. Translated by B. P. Pratten. In *Ante-Nicene Fathers*, edited by Alexander Roberts, James Donaldson, and A. Cleveland Coxe, 2:280–333. Buffalo, N.Y.: Christian Literature, 1885.
———. *On the Resurrection*. Translated by B. P. Pratten. In Roberts, Donaldson, and Coxe, *Ante-Nicene Fathers*, 2:334–67. Buffalo, N.Y.: Christian Literature, 1885.
Augustine. *Augustine on Genesis: Two Books on Genesis against the Manichees and the Literal Interpretation of Genesis, an Unfinished Book*. Translated by Roland J. Teske. Washington, D.C.: The Catholic University of America Press, 1991.
———. *City of God*. Translated by Henry Bettenson. New York: Penguin, 2003.
———. *Confessions*. Translated by Henry Chadwick. Oxford: Oxford University Press, 1991.
———. *Genesi ad Literam (On the Literal Meaning of Genesis)*. Translated by John Hammond Taylor. Mahwah, N.J.: Paulist Press, 1982.
———. *Tractates on the Gospel of John*. Translated by John Gibb. In *Nicene and Post-Nicene Fathers*, edited by Philip Schaff, Series 1, 7:4–776. Buffalo, N.Y.: Christian Literature, 1888.
———. *On the Trinity*. In *Nicene and Post-Nicene Fathers*, edited by Philip Schaff, Series 1, 3:21–475. London: T & T Clark, 1887.
———. *The Works of Saint Augustine: A Translation for the 21st Century; Sermons*. Translated by Edmund Hill. Hyde Park, N.Y.: New City Press, 1990.
Babich, Babette. "The Philosopher and the Volcano: On the Antique Sources of Nietzsche's Übermensch." *Philosophy Today* 55, Supplement (2011): 206–24.
Bammel, Caroline P. Hammond. "Die fehlehnden Bände des Römerbriefkommentars des Origenes." In *Origeniana Quarta*, edited by Lothar Lies. Innsbruck: Tyrolia, 1987.
Bandt, Christoph. "Introduction to Fractals." In *Fractals, Wavelets, and Their Applications*, edited by Christoph Bandt, 3–18. New York: Springer International, 2014.
Barad, Karen. "Nature's Queer Performativity." *Qui Parle* 19, no. 2 (2011): 121–58.
Bartholomew, Craig, and Ryan P. O'Dowd. *Old Testament Wisdom Literature: A Theological Introduction*. Downer's Grove, Ill.: InterVarsity Press Academic, 2018.
bar Yochai, Shimon. *Resurrection and Reincarnation*. Translated by Shabtai Teicher. Accessed March 27, 2021, https://www.chabad.org/kabbalah/article_cdo/aid/380668/jewish/Resurrection-and-Reincarnation.htm.

———. "The Secret of Servitude." Translated by Moshe Miller. Accessed March 27, 2021, https://www.chabad.org/kabbalah/article_cdo/aid/380542/jewish/The-Secret-of-Servitude.htm.
Benish, Abraham. *Jewish School and Family Bible*. Vol. 1, *Pentateuch*. London: Longman, Brown, Green, and Longman's, 1852.
Bennett, Jane. *The Enchantment of Modern Life: Attachments, Crossings, and Ethics*. Princeton, N.J.: Princeton University Press, 2001.
———. "From Nature to Matter." In *Second Nature: Rethinking the Natural through Politics*, edited by Crina Archer, Laura Ephraim, and Lida Maxwell, 149–60. New York: Fordham University Press, 2013.
———. *Vibrant Matter: A Political Ecology of Things*. Durham, N.C.: Duke University Press, 2010.
———. "A Vitalist Stopover on the Way to New Materialism." In *New Materialisms: Ontology, Agency, and Politics*, edited by Diana Coole and Samantha Frost, location 589–902. Durham, N.C.: Duke University Press, 2010.
Bergant, Dianne. "Creation Theology in the Book of Job." *America: The Jesuit Review* (2008): unpaginated.
Berkeley, George. *Three Dialogues between Hylas and Philonous* (1713). Edited by Robert M. Adams. Indianapolis: Hackett, 1979.
Blanchot, Maurice. *The Infinite Conversation*. Translated by Susan Hanson. Minneapolis: University of Minnesota Press, 1992.
———. *The Writing of the Disaster*. Translated by Ann Smock. Lincoln: University of Nebraska Press, 2015.
Brakke, David. "The Body as/at the Boundary of Gnosis." *Journal of Early Christian Studies* 17, no. 2 (2009): 195–215.
———. *The Gnostics: Myth, Ritual, and Diversity in Early Christianity*. Cambridge, Mass.: Harvard University Press, 2010.
Bristow, William. "Enlightenment." *Stanford Encyclopedia of Philosophy* (2017). https://plato.stanford.edu/entries/enlightenment/.
Brons, David. "Christ and the Church." Accessed March 27, 2021. http://gnosis.org/library/valentinus/Christ_and_Church.htm.
———. "The Name and Naming in Valentinianism." Accessed March 27, 2021. http://gnosis.org/library/valentinus/Name_Naming.htm.
———. "Valentinian Theology." Accessed March 27, 2021. http://gnosis.org/library/valentinus/Valentinian_Theology.htm.
Brown, Peter. *Augustine of Hippo: A Biography*. Berkeley: University of California Press, 1967.
Bryant, Levi. *The Democracy of Objects*. Ann Arbor, Mich.: Open Humanities, 2011.
Buckley, Jorunn Jacobsen. "A Cult-Mystery in *the Gospel of Philip*." *Journal of Biblical Literature* 99, no. 4 (1980): 569–81.
Burkert, Walter. *Lore and Science in Ancient Pythagoreanism*. Cambridge, Mass.: Harvard University Press, 1972.

Burrus, Virginia. *Ancient Christian Ecopoetics: Cosmologies, Saints, Things.* Philadelphia: University of Pennsylvania Press, 2018.
Bynum, Carolyn Walker. *The Resurrection of the Body in Western Christianity 200–1336.* New York: Columbia University Press, 1995.
Callender, Dexter E. Jr., ed. *Myth and Scripture: Contemporary Perspectives on Religion, Language, and Imagination.* Atlanta: Society for Biblical Literature, 2014.
———. "Myth and Scripture: Dissonance and Convergence." In *Myth and Scripture: Contemporary Perspectives on Religion, Language, and Imagination,* edited by Dexter E. Callender Jr., 27–50. Atlanta: Society for Biblical Literature, 2014.
Carson, Anne. *Eros the Bittersweet: An Essay.* Princeton, N.J.: Princeton University Press, 1986.
Cary, Philip. *Inner Grace: Augustine in the Traditions of Plato and Paul.* Oxford: Oxford University Press, 2008.
Cassian, John. *The Conferences.* Translated by Boniface Ramsey. Mahwah, N.J.: Paulist Press, 1997.
Charles, Robert Henry, ed. *The Apocrypha and Pseudepigrapha of the Old Testament in English.* Oxford: Clarendon, 1913.
Chin, Catherine Michael. "Cosmos." In *Late Ancient Knowing: Explorations in Intellectual History,* edited by Catherine Michael Chin and Moulie Vidas, 99–116. Berkeley: University of California Press, 2015.
Clement of Alexandria. *Excerpts of Theodotus (Excerpta Ex Theodoto).* Translated by Robert Pierce Casey. London: Christophers, 1934.
Cochran, Elizabeth Agnew. "Virtuous Assent and Christian Faith: Retrieving Stoic Virtue Theory for Christian Ethics." *Journal of the Society of Christian Ethics* 30, no. 1 (2010): 117–40.
Coole, Diana. "The Inertia of Matter and the Generativity of Flesh." In Coole and Frost, *New Materialisms: Ontology, Agency, and Politics,* location 1198–1500.
Coole, Diana, and Samantha Frost, eds. *New Materialisms: Ontology, Agency, Politics.* Durham, N.C.: Duke University Press, 2010.
Cossins, Daniel. "Why We Like to Know Useless Stuff." *New Scientist,* 2017.
Cowley, A. E., and A. Neubauer. *The Original Hebrew of a Portion of Ecclesiasticus.* Oxford: Clarendon, 1897.
Cox, Ronald. *By the Same Word: Creation and Salvation in Hellenistic Judaism and Early Christianity.* Berlin: Walter de Gruyter, 2007.
Curd, Patricia. "Presocratic Philosophy." In *Stanford Encyclopedia of Philosophy,* 2016. Accessed March 27, 2021, at https://plato.stanford.edu/entries/presocratics/.
———. *A Presocratics Reader.* Indianapolis: Hackett, 2011.
Curd, Patricia, and Daniel W. Graham, eds. *The Oxford Handbook of Presocratic Philosophy.* Oxford: Oxford University Press, 2008.
Dennis, Geoffrey W. "Adam Kadmon." In *The Encyclopedia of Jewish Myth, Magic and Mysticism.* Woodbury, Minn.: Llewelyn Worldwide, 2016.
Descartes, René. *Meditations on First Philosophy.* Translated by Donald Cress. Indianapolis: Hackett, 1993.

———. *The Philosophical Writings of Descartes*. Vol. 1, *Principles of Philosophy*. Translated by R. Stoothoff, J. Cottingham, and D. Murdoch. Cambridge: Cambridge University Press, 1985.
Dillon, John. "Solomon Ibn Gabirol's Doctrine of Intelligible Matter." In *Neoplatonism and Jewish Thought*, edited by Lenn E. Goodman, 43–60. Albany: State University of New York Press, 1992.
Diogenes Laërtius. *Lives and Opinions of Eminent Philosophers, Including the Biographies of the Cynics and the Life of Epicurus*. Translated by Robert Drew Hicks. New York: Gottfried and Fritz, 2014.
Dirac, Paul. "The Quantum Theory of the Electron." *Proceedings of the Royal Society of London: Series A, Containing Papers of a Mathematical and Physical Character* 117, no. 778 (1928): 610–24.
Dunderberg, Ismo. "Recognizing the Valentinians—Now and Then." In *The Other Side: Apocryphal Perspectives on Ancient Christian "Orthodoxies,"* edited by Tobias Niklas, Candida R. Moss, Christopher Tuckett, and Joseph Verhayden, 30–53. Göttingen: Vanderhoeck and Ruprecht, 2017.
———. "Valentinus and His School." *Revista Catalana de Teologia* 37, no. 1 (2012): 131–51.
Meister Eckhart. *Meister Eckhart: Selected Writings*. Translated by Oliver Davies. New York: Penguin, 1994.
Edwards, Mark J. "The Epistle to Rheginus: Valentinianism in the Fourth Century." *Novum Testamentum* 3, no. 1 (January 1995): 76–91.
Ehrman, Bart D. "Christianity Turned on Its Head: The Alternative Vision of the Gospel of Judas." In *The Gospel of Judas from Codex Tchacos*, ed. Rodolphe Kasser et al., 77–120. Washington, D.C.: National Geographic, 2006.
Eliot, T. S. *Collected Poems 1909–1962*. New York: Harcourt, Brace, 1991.
Epicurus. *The Essential Epicurus: Letters, Principal Doctrines, Vatican Sayings, and Fragments*. Translated by Eugene O'Connor. Buffalo, N.Y.: Prometheus, 1993.
Filoramo, Giovanni. *A History of Gnosticism*. Translated by Anthony Alcock. Oxford: Basil Blackwell, 1990.
Fine, Lawrence. "Tikkun: A Lurianic Motif in Contemporary Jewish Thought." In *From Ancient Israel to Modern Judaism: Essays in Honor of Marvin Fox*, edited by Jacob Neusner, Ernest S. Frerichs, and Nahum M. Sarna, 35–54. Providence, R.I.: Brown University Press, 1989.
Flavius Josephus. *The Genuine Works of Flavius Josephus the Jewish Historian*. London: 1737.
Foucault, Michel. *The Hermeneutics of the Subject: Lectures at the Collège de France 1981–1982*. Translated by Graham Burchell. New York: Palgrave Macmillan, 2005.
Fowler, Robert. "Mythos and Logos." *Journal of Hellenic Studies* 131 (2011): 45–66.
Franke, William. "Poetry, Prophecy, and Theological Revelation." In *Oxford Research Encyclopedia in Religion* (Online), edited by John Barton. Oxford: Oxford University Press, 2016.

Franzmann, Majella. "A Complete History of Early Christianity: Taking the 'Heretics' Seriously." *Journal of Religious History* 29, no. 2 (2005): 117–28.
Freud, Sigmund. *Instincts and their Vicissitudes* (1915). Translated by James Strachey. In the Standard Edition, 14:111–40. London: Hogarth, 1957.
———. *Three Essays on the Theory of Sexuality* (1905). Translated by James Strachey. In the Standard Edition, 7:124–245. London: Hogarth, 1953.
Gaiser, Konrad. "Plato's Enigmatic Lecture 'On the Good.'" *Phronesis* 25, no. 1 (1980): 5–37.
Garber, Marjorie. *The Muses on Their Lunch Hour*. New York: Fordham University Press, 2017.
Gardet, L. "Ḳiyāma." In *Encyclopaedia of Islam*, 2nd ed, edited by P. Bearman, Th. Bianquis, C. E. Bosworth, E. van Donzel, and W. P. Heinrichs. Accessed March 27, 2021, at https://referenceworks.brillonline.com/browse/encyclopaedia-of-islam-2,
Gilbert, Maurice, S. J. "*Pirqé Avot* and Wisdom Tradition." In *Tracing Sapiential Traditions in Ancient Judaism*, edited by Jean-Sebastien Rey, Hindy Najman, and Eibert Tigchelaar, 155–71. Leiden: Brill, 2016.
Glancy, Jennifer. *Corporeal Knowledge: Early Christian Bodies*. Oxford: Oxford University Press, 2010.
Goff, Matthew. "Searching for Wisdom in and Beyond 4*Q*instruction." In Rey, Najman, and Tigchelaar, *Tracing Sapiential Traditions in Ancient Judaism*, 119–37.
Goodman, Lenn E. "Ibn Masarrah." In *History of Islamic Philosophy*, edited by Seyyed Hossein Nasr and Oliver Leaman, 277–93. London: Routledge, 1996.
———, ed. *Neoplatonism and Jewish Thought*. Albany: State University of New York Press, 1992.
Gordley, Matthew. *The Colossian Hymn in Context: An Exegesis in Light of Jewish and Greco-Roman Hymnic and Epistolary Conventions*. Tübingen: Mohr Siebeck, 2007.
Gregory of Nyssa. "On the Soul and Resurrection." In *Nicene and Post-Nicene Fathers*, edited by Philip Schaff, Series 2, 5:429–67. Grand Rapids, Mich.: William B. Eerdmans, 1888.
Guthrie, W. K. C. *A History of Greek Philosophy*. Vol. 1, *The Earlier Presocratics and the Pythagoreans*. Cambridge: Cambridge University Press, 1962.
———. *A History of Greek Philosophy*. Vol. 5, *Later Plato and the Academy*. Cambridge: Cambridge University Press, 1978.
Halevi, Judah. *Sefer Kuzari*. Translated by Hartwig Hirschfeld. Accessed March 27, 2021, https://www.sefaria.org/Sefer_Kuzari?lang=bi.
Harman, Graham. "On Vicarious Causation." In *Collapse 2: Speculative Realism*, edited by Robin Mackay, 171–206. Falmouth, UK: Urbanomic, 2007.
———. "The Road to Objects." *Continent* 3, no. 1 (2011): 171–79.
Harrison, Robert Pogue. *Gardens: An Essay on the Human Condition*. Chicago: University of Chicago Press, 2008.
Hazard, Sonia. "The Material Turn in the Study of Religion." *Religion and Society: Advances in Research* 4 (2013): 58–78.

Heidegger, Martin. *Introduction to Metaphysics*. Translated by Gregory Fried and Richard Polt. New Haven, Conn.: Yale University Press, 2000.
Hesiod. *Theogony*. Translated by Hugh G. Evelyn-White. London: William Heinemann, 1914.
Hippolytus. *The Refutation of All Heresies*. Translated by J. H. MacMahon. In *Ante-Nicene Fathers*, edited by Alexander Roberts, James Donaldson, and A. Cleveland Coxe, 5:14–405. Buffalo, N.Y.: Christian Literature, 1886.
Holland, Frederic May. *The Rise of Intellectual Liberty from Thales to Copernicus*. New York: H. Holt, 1885.
Homer. *The Iliad*. Translated by Robert Fagles. New York: Penguin, 1998.
———. *The Odyssey*. Translated by A. T. Murray. London: William Heinemann, 1919.
Huffer, Lynne. *Mad for Foucault: Rethinking the Foundations of Queer Theory*. New York: Columbia University Press, 2009.
Hyman, Arthur. "From What Is One and Simple Only What Is One and Simple Can Come to Be." In Goodman, *Neoplatonism and Jewish Thought*, 111–35.
Iamblichus. "The Life of Pythagoras." In *The Pythagorean Source Book and Library*, compiled and translated by Kenneth Sylvan Guthrie, edited by David Fideler, 1–122. Grand Rapids, Mich.: Phanes Press, 1987.
Ibn Ezra, Abraham. *Ibn Ezra on Genesis*. Accessed March 27, 2021, https://www.sefaria.org/Ibn_Ezra_on_Genesis.1.1?ven=Sefaria_Community_Translation&lang=bi.
Irenaeus. *Against Heresies*. In *Ante-Nicene Fathers*, edited by Alexander Roberts, James Donaldson, and A. Cleveland Coxe, 1:841–1450. Buffalo, N.Y.: Christian Literature, 1885.
Isaac the Blind. *Commentary on the Sefer Yisirah*. Manuscript. Cincinnati: Hebrew Union College.
Jerome. "Commentary on Ephesians 4:4." In *History of Opinions on the Scriptural Doctrine of Retribution*, edited by Edward Beecher. New York: D. Appleton, 1878.
Jonas, Hans. *The Gnostic Religion: The Message of the Alien God and the Beginnings of Christianity*. Boston: Beacon Press, 1963.
Jones, Tamsin. "Introduction: New Materialism and the Study of Religion." In *Religious Experience and New Materialism: Movement Matters*, edited by Joerg Rieger and Edward Waggoner, 1–23. London: Palgrave Macmillan, 2016.
Kaars, Eric. Thread. Accessed March 27, 2021, at https://twitter.com/erik_kaars/status/1139106297339686912.
Kalvesmaki, Joel. *The Theology of Arithmetic: Number Symbolism in Platonism and Early Christianity*. Washington, D.C.: Center for Hellenic Studies, 2013.
Kant, Immanuel. "An Answer to the Question: What Is Enlightenment?" Translated by Ted Humphrey. In *Perpetual Peace and Other Essays*, 41–48. Indianapolis: Hackett, 1983.
Kaplan, Aryeh. *Immortality, Resurrection, and the Age of the Universe: A Kabbalistic View*. Brooklyn, N.Y.: KTAV, 1993.
Keller, Catherine. *Cloud of the Impossible: Negative Theology and Planetary Entanglement*. New York: Columbia University Press, 2015.

———. *Intercarnations: Exercises in Theological Possibility.* New York: Fordham University Press, 2017.
Keller, Catherine, and Mary-Jane Rubenstein, eds. *Entangled Worlds: Religion, Science, and New Materialisms.* New York: Fordham University Press, 2017.
Kelly, J. N. D. *Jerome: His Life, Writings, and Controversies.* Peabody, Mass.: Hendrickson, 1998.
Kenny, John P. "The Platonism of the Tripartite Tractate (Nh I.5)." In *Neoplatonism and Gnosticism,* edited by Richard T. Wallis and Jay Bregman, 187–206. Albany: State University of New York Press, 1992.
King, Karen. *The Secret Revelation of John.* Cambridge, Mass.: Harvard University Press, 2009.
Klossowski, Pierre. "Nietzsche's Experience of the Eternal Return." In *The New Nietzsche,* edited by David B. Allison, 107–20. New York: Dell, 1977.
Knibb, Michael Anthony. *Essays on the Book of Enoch and Other Early Jewish Texts and Traditions.* Leiden: Brill, 2009.
Kohler, Kaufmann, and Isaac Broydé. "Empedocles of Agrigentum." Accessed March 27, 2021, at http://www.jewishencyclopedia.com/articles/5740-empedocles-of-agrigentum.
Kojève, Alexandre. *Introduction to the Reading of Hegel: Lectures on the Phenomenology of Spirit.* Translated by James H. Nichols Jr. Ithaca, N.Y.: Cornell University Press, 1980.
Kraftchick, Steven J. "Recast, Reclaim, Reject: Myth and Validity," In Callender, *Myth and Scripture,* 179–200.
Kramer, Hans Joachim. *Plato and the Foundations of Metaphysics: A Work on the Theory of the Principles and Unwritten Doctrines of Plato with a Collection of the Fundamental Documents.* Albany: State University of New York Press, 1990.
Lacan, Jacques. *The Seminar of Jacques Lacan: The Ego in Freud's Theory and in the Techniques of Psychoanalysis, 1954–55.* Translated by Sylvana Tomaselli. New York: W. W. Norton, 1988.
———. *The Seminar of Jacques Lacan: Four Fundamental Concepts of Psychoanalysis.* Translated by Jacques-Alain Miller. New York: W. W. Norton, 1998.
Laitman, Michael. *Unlocking the Zohar.* Toronto: Laitman Kabbalah, 2011.
Lambert, Wilfred G. *Babylonian Wisdom Literature.* Philadelphia: Eisenbrauns, 1996.
Latour, Bruno. "On Actor-Network Theory: A Few Clarifications." *Soziale Welt* 47 (1996): 369–81.
———. *On the Modern Cult of the Factish Gods.* Durham, N.C.: Duke University Press, 2010.
———. *Pandora's Hope: Essays on the Reality of Science Studies.* Cambridge, Mass.: Harvard University Press, 1999.
———. "On Recalling ANT." *Sociological Review* 47, no. 1 (1999): 15–25.
Lupieri, Edmondo. *The Mandaeans: The Last Gnostics.* Translated by Charles Hindley. Cambridge: Eerdmans, 2002.
Luria, Isaac. *Collected Works on Kabbalah.* Jerusalem: No publisher given, 1988.

———. *Gate of Reincarnations*. Recorded by Chaim Vital, translated by Yitzchok bar Chaim. Accessed March 27, 2021, at https://www.chabad.org/kabbalah/article_cdo/aid/378771/jewish/Gate-of-Reincarnations.htm.

MacKendrick, Karmen. *Divine Enticement: Theological Seductions*. New York: Fordham University Press, 2012.

Maimonides (Moses ben Maimon). *Commentary on Mishnah Sanhedrin*. Translated by Sefaria Community. 1158. Accessed March 27, 2021, at https://www.sefaria.org/Rambam_on_Mishnah_Sanhedrin?lang=bi.

———. *Mishnah Torah, Law of Kings and Wars*. Translated by Reuven Brauner. 2012. Accessed March 27, 2021, at https://www.sefaria.org/Mishneh_Torah%2C_Kings_and_Wars.1?ven=Laws_of_Kings_and_Wars._trans._Reuven_Brauner,_2012&lang=bi.

———. *Mishnah Torah, Repentance*. Translated by Simon Glazer. NP: Maimonides, 1927. Accessed March 27, 2021, at https://www.sefaria.org/Mishneh_Torah_Repentance?lang=bi.

———. "The 13 Principles and the Resurrection of the Dead." In *Medieval Sourcebook*, edited by Paul Halsall. Revised 2019. Accessed March 27, 2021, at https://sourcebooks.fordham.edu/source/rambam13.asp.

———. *Moses Maimonides' Treatise on the Resurrection*. Translated by Fred Rosner. Lanham Md.: Rowman and Littlefield, 2004.

Malter, Henry. *Saadia Gaon: His Life and Works*. Philadelphia: Jewish Publication Society of America, 1921.

Mandelbrot, Benoit. *The Fractal Geometry of Nature*. New York: W. H. Freeman, 1982.

Mansfeld, Jaap. "Aristotle and Others on Thales, or the Beginnings of Natural Philosophy." *Mnemosyne* 38, no. 1–2 (1985): 109–29.

Markschies, Christoph. *Gnosis: An Introduction*. Translated by John Bowen. London: T & T Clark, 2003.

Martin, Dale B. *The Corinthian Body*. New Haven, Conn.: Yale University Press, 1995.

McGlothlin, Thomas Dwight. "Raised to Newness of Life: Resurrection and Moral Transformation in Second- and Third-Century Christian Theology." Ph.D. diss., Duke University, 2015.

Merleau-Ponty, Maurice. *The Visible and the Invisible*. Translated by Alphonso Lingis. Evanston, Ill.: Northwestern University Press, 1968.

Meyer, Marvin. "Gnosticism, Gnostics, and the Gnostic Bible." In *The Gnostic Bible*, edited by Willis Barnstone and Marvin W. Meyer, 1–20. Boston: Shambala, 2003.

———, ed. *The Nag Hammadi Scriptures*. New York: HarperCollins, 2008.

Miles, Margaret. *Desire and Delight: A New Reading of Augustine's Confessions*. New York: Crossroad, 1992.

Miller, Mitchell. "The Choice between the Dialogues and the 'Unwritten Teachings': A Scylla and Charybdis for the Interpreter?" In *The Third Way: New*

Directions in Platonic Studies, edited by Francisco Gonzalez, 225–44. Lanham, Md.: Rowman and Littlefield, 1995.

Miller, Patricia Cox. "Adam, Eve, and the Elephants: Asceticism and Animality in Late Ancient Christianity." In *Ascetic Culture: Essays in Honor of Philip Rousseau*, edited by Blake Leyerle and Robin Young, 253–68. Notre Dame, Ind.: Notre Dame University Press, 2013.

———. *In the Eye of the Animal: Zoological Imagination in Ancient Christianity*. Philadelphia: University of Pennsylvania Press, 2018.

———. "'Plenty Sleeps There': The Myth of Eros and Psyche in Plotinus and Gnosticism." In Wallis and Bregman, *Neoplatonism and Gnosticism*, 221–38.

———. "'Words With an Alien Voice': Gnostics, Scripture, and Canon." *Journal of the American Academy of Religion* 57, no. 3 (1989): 459–83.

Moffatt, James. "Pistis Sophia." In *Encyclopaedia of Religion and Ethics*, edited by James Hastings, 45–47. New York: Charles Scribner's Sons, 1919.

Moll, Sebastian. *The Arch-Heretic Marcion*. Tübingen: Mohr Siebeck, 2010.

Moore, Edward. "Gnosticism." In *Internet Encyclopedia of Philosophy*. Accessed March 27, 2021, at http://www.iep.utm.edu/gnostic/.

Morton, Timothy. *The Ecological Thought*. Cambridge, Mass.: Harvard University Press, 2012.

———. "Here Comes Everything: The Promise of Object-Oriented Ontology." *Qui Parle* 19, no. 2 (Spring/Summer 2011): 163–90.

Nachmanides (Moses ben Nachman). *Commentary on Genesis (Ramban on Genesis)*. Accessed March 27, 2021, https://www.sefaria.org/Ramban_on_Genesis?lang=bi.

———. *Sh'ar ha-Gemul*, as *Sh'ar Ha'Gemul of the Ramban*. Translated by Sefaria Community. Accessed March 27, 2021, at https://www.sefaria.org/Sha'ar_Ha'Gemul_of_the_Ramban?lang=bi.

Nancy, Jean-Luc. *The Pleasure in Drawing*. Translated by Philip Armstrong. New York: Fordham University Press, 2013.

Nemesius of Emesa. *On the Nature of Man*. In *The Hellenistic Philosophers*, vol. 1, *Translations of the Principle Sources, with Philosophical Commentary*. Edited and translated by A. A. Long and D. N. Sedley. Cambridge: Cambridge University Press, 1987.

Nietzsche, Friedrich. *Ecce Homo: How One Becomes What One Is*. Translated by Walter Kaufmann and R. J. Hollingdale. New York: Penguin Classics, 1992.

———. *The Gay Science*. Translated by Walter Kaufmann. New York: Vintage, 1974.

———. *Philosophy in the Tragic Age of the Greeks*. Translated by Marianne Cowan. Washington, D.C.: Regnery, 1962.

———. *The Selected Letters of Friedrich Nietzsche*. Edited by Christopher Middleton. Indianapolis: Hackett, 1969.

———. *Untimely Meditations*. Translated by R. J. Hollingdale. Cambridge: Cambridge University Press, 1997.

———. *The Will to Power*. Translated by Walter Kaufmann and R. J. Hollingdale. New York: Vintage, 1968.

O'Grady, Patricia F. *Thales of Miletus: The Beginnings of Western Science and Philosophy*. New York: Routledge, 2016.
O'Hara, Frank. "Meditations in an Emergency." In *Meditations in an Emergency*. New York: Grove, 1957.
Olsen, Roger E. *The Story of Christian Theology: Twenty Centuries of Tradition and Reform*. Downer's Grove, Ill.: InterVarsity Press, 1999.
Orfanos, Spyros D. "Mythos and Logos." *Psychoanalytic Dialogues* 16, no. 4 (2006): 481–99.
Origen. *Against Celsus*. Translated by Frederick Cromble. In *Ante-Nicene Fathers*, edited by Alexander Roberts, James Donaldson, and A. Cleveland Coxe, 4:848–1570. Buffalo, N.Y.: Christian Literature, 1885.
———. *Commentary on the Gospel according to John*. Books 1 to 10. Translated by Ronald E. Heine. Washington, D.C.: The Catholic University of America Press, 1989.
———. *Commentary on Matthew*. In *Origenes Werke*, edited by E. Benz and E. Klostermann. Vol. 10. Leipzig: Hinrichs, 1935.
———. *On First Principles*. Translated by G. W. Butterworth. Gloucester, Mass.: Peter Smith, 1973.
———. *On Prayer*. Translated by William A. Curtis. Grand Rapids, Mich.: Christian Classics Ethereal Library, 2001.
Pagels, Elaine. *The Gnostic Paul: Gnostic Exegesis of the Pauline Letters*. Philadelphia: Fortress, 1975.
Parmenides. *On Nature (Peri Physeos)*. Translated by David Gallop, modified from Allan Randall, Richard D. McKirahan Jr., Jonathan Barnes, John Manley Robinson, et al. 1996. Accessed March 27, 2021, http://www.allanrandall.ca/Parmenides.html.
Partenie, Catalin. "Introduction." In *Plato's Myths*, edited by Catalin Partenie, 1–27. Cambridge: Cambridge University Press, 2009.
———. "Plato's Myths." *Stanford Encyclopedia of Philosophy* (2018). Accessed March 27, 2021, https://plato.stanford.edu/entries/plato-myths.
Patai, Raphael. *Man and Temple: In Ancient Jewish Myth and Ritual*. London: Thomas Nelson and Sons, 1947.
Pearson, Birger. *Ancient Gnosticism: Traditions and Literature*. Minneapolis: Fortress, 2007.
Peirce, Charles Sanders. *Collected Papers of Charles Sanders Peirce*. Vol. 1, *Principles of Philosophy*; Vol. 2, *Elements of Logic*. Cambridge, Mass.: Harvard University Press, 1932.
Pepple, John. "The Unwritten Doctrines: Plato's Answer to Speusippus." 1997. http://personal.kenyon.edu/pepplej/#section%20VII. Accessed March 27, 2021.
Pessin, Sarah. *Ibn Gabirol's Theology of Desire: Matter and Method in Jewish Medieval Neoplatonism*. Cambridge: Cambridge University Press, 2013.
Philo. *The Works of Philo: Complete and Unabridged*. Translated by Charles D. Yonge. London: H. G. Bohn, 1854–90.

Plato. *Apology*. Translated by Harold North Fowler. London: William Heinemann, 1966.
———. *Parmenides*. Translated by Harold North Fowler. London: William Heinemann, 1925.
———. *Phaedo*. Translated by G. M. A. Grube. Revised by John M. Cooper. In *Plato: Five Dialogues: Euthyphro, Apology, Crito, Meno, Phaedo*. Indianapolis: Hackett, 2002.
———. *Phaedrus*. Translated by Alexander Nehamas and Paul Woodruff. Indianapolis: Hackett, 1995.
———. *Philebus*. Translated by Dorothea Frede. Indianapolis: Hackett, 1993.
———. *Republic*. Translated by Robin Waterfield. Oxford: Oxford University Press, 2008.
———. *Sophist*. Translated by Harold N. Fowler. London: William Heinemann, 1921.
———. *Statesman*. Translated by Eva Brann, Peter Kalkavage, and Eric Salem. Indianapolis: Hackett, 2012.
———. *Symposium*. Translated by Alexander Nehamas and Paul Woodruff. Indianapolis: Hackett, 1989.
———. *Theaetetus*. Translated by John McDowell. Oxford: Oxford University Press, 2014.
———. *Timaeus*. Translated by W. R. M. Lamb. London: William Heinemann, 1925.
Plotinus. *Plotinus: The Enneads*. Translated by Stephen MacKenna. New York: Penguin, 1991.
Porphyry. *The Life of Pythagoras*. In Guthrie and Fideler, *The Pythagorean Sourcebook and Library*, 123–36.
Rankin, David. *The Early Church and the Afterlife: Post-Death Existence in Athenagoras, Tertullian, Origen, and the Letter to Rheginos*. New York: Routledge, 2017.
Rashi (Rabbi Shlomo Yitzchaki). *Pentateuch with Rashi's Commentary*. Translated by M. Rosenbaum and A. M. Silbermann. Accessed March 27, 2021, https://www.sefaria.org/Rashi_on_Genesis?lang=bi.
Rey, Jean-Sebastien, Hindy Najman, and Eibert Tigchelaar, eds. *Tracing Sapiential Traditions in Ancient Judaism*. Leiden: Brill, 2016.
Riedweg, Christoph. *Pythagoras: His Life, Teaching, and Influence*. Translated by Steven Rendall. Ithaca, N.Y.: Cornell University Press, 2005.
Robinson, James, ed. *The Nag Hammadi Library*. New York: HarperCollins, 1990.
Rojcewicz, Richard. "Everything Is Water." *Research in Phenomenology* 44 (2014): 194–211.
Rojtman, Betty. *Black Fire on White Fire: An Essay on Jewish Hermeneutics, from Midrash to Kabbalah*. Translated by Steven Rendall. Berkeley: University of California Press, 1998.
Rosen-Zvi, Ishay. "The Wisdom Tradition in Rabbinic Literature and Mishnah Avot." In Rey, Najman, and Tigchelaar, *Tracing Sapiential Traditions in Ancient Judaism*, 172–90.

Rowe, C. J. "Myth, History, and Dialectic in Plato's Republic and Timaeus-Critias." In *From Myth to Reason? Studies in the Development of Greek Thought*, edited by R. Buxton, 251–62. Oxford: Oxford University Press, 1999.
Rubenstein, Mary-Jane. *Pantheologies: Gods, Worlds, Monsters*. New York: Columbia University Press, 2018.
———. *Worlds without End: The Many Lives of the Multiverse*. New York: Columbia University Press, 2014.
Saadia Gaon. "Treatise 7, Concerning the Resurrection of the Dead in This World." In *The Book of Beliefs and Opinions*, trans. Samuel Rosenblatt, 264–85. New Haven, Conn.: Yale University Press, 1976.
———. רס"ג ופירוש תרגום עם אנטיוכס ומגילת דניאל ספר. Edited by Y. Qafih. Jerusalem: Committee for the Publication of the Books of R. Sa'adia Gaon, 1981.
Sayre, Kenneth. *Plato's Late Ontology: A Riddle Resolved*. Princeton, N.J.: Princeton University Press, 1983.
Schlossberg, Eliezer, and Dov Schwartz. "From Periphery to Center: Early Discussion of Resurrection in Medieval Jewish Thought." *Hebrew Union College Annual* 89 (2018): 177–96.
Schneider, Stanley, and Morgan Seelenfreund. "Kotnot Or (Genesis 3:21): Skin, Leather, Light, or Blind?" *Jewish Bible Quarterly* 40, no. 2 (2012): 116–25.
Schutte, Heinz R. *Weltseele: Geschichte und Hermeneutik*. Frankfurt: J. Knecht, 1993.
Schwartz, Howard. *A Palace of Pearls: The Stories of Rabbi Nachman of Bratslav*. Oxford: Oxford University Press, 2018.
Scott, Alan. *Origen and the Life of the Stars: A History of an Idea*. Oxford: Oxford University Press, 1991.
Shaviro, Steven. *Posthumanities: Universe of Things: On Speculative Realism*. Minneapolis: University of Minnesota Press, 2014.
Shaw, Gregory. "Theurgy and the Platonist's Luminous Body." In *Practicing Gnosis: Ritual, Magic, Theurgy, and Liturgy in Nag Hammadi, Manichaean and Other Ancient Literature; Essays in Honor of Birger A. Pearson*, edited by Gregory Shaw, April DeConick, and John D. Turner, 537–56. Leiden: Brill, 2013.
———. *Theurgy and the Soul: The Neoplatonism of Iamblichus*. University Park: University of Pennsylvania Press, 2003.
Shenker, Orly R. "Fractal Geometry Is Not the Geometry of Nature." *Studies in the History and Philosophy of Science* 25, no. 6 (1994): 967–81.
Silver, Daniel J. "Nachmanides' Commentary on the Book of Job." *Jewish Quarterly Review* 60, no. 1 (1969): 9–26.
———. "The Resurrection Debate." In *Moses Maimonides' Treatise on Resurrection*, edited by Fred Rosner, 71–102. Northvale, N.J.: Jason Aronson, 1997.
Stead, G. C. "The Valentinian Myth of Sophia." *Journal of Theological Studies* 11, no. 1 (1969): 75–104.
Tatian. "Address to the Greeks." In *Ante-Nicene Fathers*, edited by James Donaldson, Alexander Roberts, and A. Cleveland Coxe, 2:98–129. Buffalo, N.Y.: Christian Literature, 1885.

Tertullian. "Against Marcion," In *Ante-Nicene Fathers*, vol. 3, edited by James Donaldson, Alexander Roberts, and A. Cleveland Coxe, 432–822. Buffalo N.Y.: Christian Literature Publishing, 1885.
———. *Against the Valentinians*. Translated by Mark T. Riley. 1971. Accessed March 27, 2021. http://tertullian.org/articles/riley_adv_val/riley_00_index.htm.
———. "De Resurrectione Carnis." In *Patrilogia Latina*, edited by J. P. Migne, 2:791–886. Paris: J. P. Migne, 1844.
———. *On the Resurrection*. Translated by Peter Holmes. In *Ante-Nicene Fathers*, edited by A. Cleveland Coxe, Alexander Roberts, and James Donaldson, 3:952–1039. Buffalo, N.Y.: Christian Literature, 1885.
———. "Prescription Against Heretics." In *Ante-Nicene Fathers*, edited by James Donaldson, Alexander Roberts, and A. Cleveland Coxe, 3:383–431. Buffalo, N.Y.: Christian Literature, 1885.
Theophrastus. *On First Principles (Known as His Metaphysics)*. Translated by D. Gutas. Leiden: Brill, 2010.
Thomas Aquinas. *Summa Contra Gentiles*. Translated by the Fathers of the English Dominican Province. New York: Benzinger Brothers, 1929.
———. *Summa Theologiae*. Translated by the Fathers of the English Dominican Province. London: Thomas Baker, 1917.
Thomassen, Einar. *The Spiritual Seed: The Church of the "Valentinians."* Leiden: Brill, 2006.
———. "The Structure of the Transcendent World in the Tripartite Tractate (NHC I, 5)." *Vigilae Chrsitianae* 34, no. 4 (1980): 358–75.
Trouillard, Jean. *Le mystagogie de Proclos*. Paris: Les Belles Lettres, 1982.
Turner, Denys. "How to be an Atheist." *New Blackfriars* 83, no. 977–978 (2002): 317–35.
Turner, John D. *Sethian Gnosticism and the Platonic Tradition*. Peeters, Belgium: Le Presses de l'Université Laval, 2006.
———. "Sethianism." In *The Encyclopedia of Ancient History*, edited by Roger S. Bagnall, Kai Brodersen, Craige B. Champion, Andrew Erskine, and Sabine R. Huebner, 6176–79. London: Blackwell, 2013.
Turner, John D., and Orval S. Wintermute, translators. "Allogenes." In Robinson, *Nag Hammadi Library*, 490–500.
Ullman, Yimiyahu. "Reincarnation." Accessed March 27, 2021, https://ohr.edu/1077.
Uusimäki, Elisa. "Spiritual Formation in Hellenistic Jewish Wisdom Teaching." In Rey, Najman, and Tigchelaar, *Tracing Sapiential Wisdom Traditions in Ancient Judaism*, 57–70.
van der Horst, Pieter W. "The Measurement of the Body: A Chapter in the History of Ancient Jewish Mysticism." In *Effigies Dei: Essays on the History of Religions*, ed. Dirk van der Plas, 56–68. Studies in the History of Religions (Supplements to Numen) 51. Leiden: Brill, 1987.
van der Toorn, Karel. "Why Wisdom Became a Secret: On Wisdom as a Written Genre." In *Wisdom Literature in Mesopotamia and Israel*, edited by Richard J. Clifford, 21–29. Atlanta: Society of Biblical Literature, 2007.

Vital, Chaim. *The Tree of Life*. Translated by E. Collé and H. Collé. NP: CreateSpace, 2015.
Von Hendy, Andrew. *The Modern Construction of Myth*. Bloomington: Indiana University Press, 2002.
Wallis, Richard T., and Jay Bregman, eds. *Neoplatonism and Gnosticism*. Albany: State University of New York Press, 1992.
White, Stephen A. "Milesian Measures: Time, Space, and Matter." In *The Oxford Handbook of Presocratic Philosophy*, edited by Patricia Curd and Daniel W. Graham, 89–122. Oxford: Oxford University Press, 2008.
Whitehead, Alfred North. *Process and Reality*. New York: Free Press, 1978.
Whitmarsh, Tim. *Battling the Gods: Atheism in the Ancient World*. New York: Vintage, 2015.
Williams, Michael A. "A Life Full of Meaning and Purpose: Demiurgical Myths and Social Implications." In *Beyond the Gnostic Gospels: Studies Building on the Work of Elaine Pagels*, edited by Eduard Iricinschi, Lance Jenott, Nicola Denzey Lewis, and Philippa Townsend, 19–59. Tübingen: Mohr Siebeck, 2013.
Wolfson, Elliot R. *Language, Eros, Being: Kabbalistic Hermeneutics and Poetic Imagination*. New York: Fordham University Press, 2005.
———. "Mystical Rationalization of the Commandments in 'Sefer ha-Rimmon.'" *Hebrew Union College Annual* 59, no. 1988 (1988): 217–51.
———. *Venturing Beyond: Law and Morality in Kabbalistic Mysticism*. Oxford: Oxford University Press, 2006.
Wright, M. R. "Presocratic Cosmologies." In Curd and Graham, *The Oxford Handbook of Presocratic Philosophy*, 414–28.
Xenophanes. *Fragments: A Text and Translation With a Commentary*. Translated by James Lesher. Toronto: University of Toronto Press, 1992.

Index

1 Enoch (apocryphal scripture; also *The Book of Enoch*), 42, 47, 123, 139, 155, 177–78
2 Baruch (also *The Book of the Apocalypse of Baruch*), 102, 122, 155, 169, 177
2 Enoch (apocryphal scripture; also *The Book of the Secrets of Enoch*), 149, 151, 177
13 *Principles and the Resurrection of the Dead, The* (Maimonides), 167
1958 Presidential Address to the Aristotelian Society (Popper), 15

Aaron (Biblical figure), 101
Abraham (Biblical figure), 151
absence 57, 78
Academy (ancient school), 4, 15, 66–67, 136
Acts of the Apostles, 94–95, 103, 166
Adam: Biblical figure, 4, 12, 39, 42, 46, 55, 111, 139, 147, 149, 154; cosmic and microcosmic figure, 46–48, 56, 93, 103, 110, 112, 154; primal human, 12, 44–46, 50, 53–54, 83, 110–12, 122
"Adam, Eve, and the Elephants: Asceticism and Animality in Late Ancient Christianity" (P.C. Miller), 136
Adam Kadmon, 46, 54
"Adam Kadmon" (Dennis), 151
Adam Qadma'ah. *See* Adam Kadmon
adornment, 45, 48–49, 80, 103, 109
Adorno, Theodor: *Dialectic of the Enlightenment: Philosophical Fragments*, 22–24, 37, 143, 146

Aeons, 40, 41, 73–79, 85–88, 90, 153, 164. *See also* All, The; Entireties; Fullness; Pleroma
aer, 17, 28, 141
afterlife, 33, 62, 97–98, 111
Against Celsus (Origen), 172, 176
Against Heresies (Irenaeus), 11, 156, 159–65, 167–69
Against Marcion (Tertullian), 169
Against the Valentinians (Tertullian), 161–63
Agamben, Giorgio, 2, 136
agency. *See* matter
Alcmaeon of Croton, 99, 142
Alexander of Aphrodisias, 66
Alexandria (Egyptian city), 4,11, 150, 161, 163–64
All, The, 72–80, 83, 85, 87–88, 107, 153. *See also* Aeons; Entireties; Fullness; Pleroma
Allogenes, 73, 154, 160
Altmann, Alexander, 51, 54; "Gnostic Themes in Rabbinic Cosmology," 153; "A Note on the Rabbinic Doctrine of Creation," 170; *Studies in Religious Philosophy and Mysticism*, 150–53
amorphousness, 17, 24, 68, 77–78. *See also* flesh: formlessness of
Anaximander, 15, 28, 141, 146, 176
Anaximenes, 28, 141
Ancient Christian Ecopoetics: Cosmologies, Saints, Things (Burrus), 8–9, 138, 173

Ancient Gnosticism: Traditions and Literature (Pearson), 159
androgyny, 40–41, 43, 49, 71, 148, 152
Andruska, Jennifer: *Wise and Foolish Love in the Song of Songs*, 137
angels, 4, 9, 40–43, 47–48, 50, 85, 94, 97, 101, 164
Anima, De. See *On the Soul (De Anima)* (Aristotle)
animation, 23–4, 27, 29, 35, 99, 109, 119, 124, 144
animism, 9, 20, 23
Ann Lee, Mother, 155
Annas, Julia: *Aristotle's Metaphysics Books M and N*, 158
anointing, 60, 106, 172, 177, 179
anthropocentrism, 2, 5–7, 11, 43, 56, 58, 130, 136
anthropomorphism, 3, 30, 129, 136
antimatter, 1, 135
"Answer to the Question, An: What Is Enlightenment?" (Kant), 142
Antiquities of the Jews, The (Flavius Josephus), 166
apeiron, 17, 28, 67, 141
Apocalypse of Adam, The, 40, 148
Apocrypha, 139, 169, 177
Apocryphon of John, The, 70, 152. See also *The Secret Book of John*
apokatastasis, 103–4, 109, 119, 170. See also restoration
Apology (Plato), 4, 136, 142, 163
Aquinas. See Thomas Aquinas
Arch-Heretic Marcion, The (Moll), 139
Aristophanes, 40, 64, 146
Aristotelianism, 9–10, 15, 24–25, 27, 66, 113, 117, 159
Aristotle 17–19, 21, 23–24, 27, 29–30, 64, 67, 127, 174; *Metaphysics*, 66, 140–41, 143–45, 157–59, 179; *On the Heavens*, 144; *On the Soul (De Anima)*, 114, 144; *Physics*, 28, 66, 141, 144, 150
"Aristotle and Others on Thales, or the Beginnings of Natural Philosophy" (Mansfeld), 141
Aristotle's Metaphysics Books M and N (Annas), 158
Aristoxenus, 67, 158
arithmetic, 64–5, 70, 88. See also mathematics; number
asceticism, 7, 11, 33, 94
astonishment, 2, 124, 136

Athanasius: *The Life of Antony and the Letter to Marcellinus*, 36, 146
atheism, 2, 29–30, 144
Athenagoras, 29, 99, 112–16, 118; *On the Resurrection*, 168, 174–75; *A Plea for the Christians*, 144
atonement, 43, 61–2, 149
Augustine of Hippo, Saint: *Augustine on Genesis: Two Books on Genesis against the Manichees and the Literal Interpretation of Genesis, an Unfinished Book*, 156; *City of God*, 97–98, 163, 167–68, 174–75, 178; *Confessions*, 127, 156, 168, 178; *Genesi ad Literam (On the Literal Meaning of Genesis)*, 156; *Tractates on the Gospel of John*, 167; *On the Trinity*, 148, 167; *The Works of Saint Augustine: A Translation for the 21st Century; Sermons*, 163
Augustine of Hippo: A Biography (Brown), 178

Babich, Babette: "The Philosopher and the Volcano: On the Antique Sources of Nietzsche's Übermensch," 176
Babylonian Wisdom Literature (Lambert), 138
Baker, David: "Big History, Critical Thinking, and Transdisciplinarity," 52
Bammel, Caroline P. Hammond: "Die fehlenden Bände des Römerbriefkommentars des Origenes," 178
Bandt, Christoph:"Introduction to Fractals," 88, 164
baptism, 81–2, 85, 172
Bar Yochai, Shimon: *Resurrection and Reincarnation*, 174; *The Secret of Servitude*, 173
Barad, Karen: "Nature's Queer Performativity," 5, 137
Barbelo (mythical figure), 70, 73, 152, 160
Barbery, Muriel: *The Elegance of the Hedgehog*, 1
Barnstone, Willis: *The Gnostic Bible*, editor, 154
Bartholomew, Craig: *Old Testament Wisdom Literature: A Theological Introduction*, 138
Battling the Gods: Atheism in the Ancient World (Whitmarsh), 19, 30, 142, 144
Bava Batra, 54, 153
Beachy-Quick, Dan: *Of Silence and Sound*, 38
beauty, 7, 19, 30, 33–34, 54, 67–68, 79–80, 82, 90–91, 103, 109–10, 114–16, 119–28, 177–78
Beginning of Wisdom, The, 171

INDEX

belief, 8, 23, 26–29, 72, 79–82, 93–96, 98, 112–13, 124, 176
Benish, Abraham: *Jewish School and Family Bible*, editor, 147
Bennett, Jane: *The Enchantment of Modern Life*, 143; "From Nature to Matter," 137; *Vibrant Matter: A Political Ecology of Things*, 143; "A Vitalist Stopover on the Way to New Materialism," 137
ben Asher, Jacob: *Tur HaAroch*, 148
ben Joseph, Saadia, Gaon. *See* Saadia Gaon
ben Maimon, Moses, Rabbi. *See* Maimonides
ben Nachman, Moses, Rabbi. *See* Nachmanides
Ben Sira. *See* Sirach
Bereishit Rabbah. *See Genesis Rabbah*
Bergant, Dianne: "Creation Theology in the Book of Job," 7, 137
Berkeley, George: *Three Dialogues between Hylas and Philonous*, 135
Bible, 4, 7, 147, 154–55. *See also* specific books
"Big History, Critical Thinking, and Transdisciplinarity" (Baker), 52
Bion, 29, 144
Black Fire on White Fire: An Essay on Jewish Hermeneutics, From Midrash to Kabbalah (Rojtman), 57, 155
Blanchot, Maurice, 37–38; *The Infinite Conversation*, 146; *The Writing of the Disaster*, 87, 147
body, 36, 49, 61,63, 77–78, 82–84, 93–95, 97–99, 103, 107–8, 110, 112–14, 116–17, 121, 123–24, 152, 164, 171, 175. *See also* embodiment; incarnation
"Body as/at the Boundary of Gnosis, The" (Brakke), 150, 152–53, 155–56, 160–63
Book of Thomas the Contender, The, 154
Books of Jeu and Untitled Treatise from the Bruce Codex, The, 153
boundary, 9–10, 12, 38–39, 45, 51, 55, 63–64, 73, 93, 127, 147. *See also* limit
boundlessness, 72, 150. *See also* apeiron
Brakke, David, 50–51, 60, 63, 74, 77, 82; "The Body as/at the Boundary of Gnosis," 150, 152–53, 155–56, 160–63; *The Gnostics: Myth, Ritual, and Diversity in Early Christianity*, 139
breath, 38–39, 42, 44–45, 57, 94–95, 99–100, 149–50, 169
brightness, 54. *See also* light; luminosity
Bristow, William: "Enlightenment," 142

Brons, David 81, 85; "Christ and the Church," 163; "The Name and Naming in Valentinianism," 164; "Valentinian Theology," 161–62, 164
Brown, Peter: *Augustine of Hippo: A Biography*, 178
Broydé, Isaac: "Empedocles of Agrigentum," 165
Bryant, Levi; *The Democracy of Objects*, 2, 136
Buckley, Jorunn Jacobsen: "A Cult-Mystery in the *Gospel of Philip*," 105, 171
burial, 43, 149
Burkert, Walter: *Lore and Science in Ancient Pythagoreanism*, 158
Burrus, Virginia: *Ancient Christian Ecopoetics: Cosmologies, Saints, Things*, 8–9, 138, 173
By the Same Word: Creation and Salvation in Hellenistic Judaism and Early Christianity (Cox), 138
Bynum, Carolyn Walker: *The Resurrection of the Body in Western Christianity 200–1336*, 177

Callender, Dexter E., Jr: "Myth and Scripture: Dissonance and Convergence," 32, 36, 145–46
cannibalism, 116–17
Carson, Anne: "Contempts," 72; *Eros the Bittersweet: An Essay*, 161
Cary, Philip: *Inner Grace: Augustine in the Traditions of Plato and Paul*, 178
Cassian. *See* John Cassian
celestial bodies, 101, 109, 113, 122, 169, 172
chaos, 18, 24, 27, 39, 43, 77–79, 88, 99, 171
Chin, Catherine Michael: "Cosmos," 9, 138
"Choice Between the Dialogues and the 'Unwritten Teachings,' The: A Scylla and Charybdis for the Interpreter?" (M. Miller), 157
Christianities, 8–9, 44, 105–6
Christianity, 4, 6, 8, 10, 12, 21, 50, 60–61, 63, 71, 81, 95–96, 102, 148, 154, 156, 178. *See also* Gnosticism; orthodoxy; Valentinianism
"Christianity Turned on its Head: The Alternative Vision of the Gospel of Judas" (Ehrman), 139
Christ, 8, 11–12, 42, 45, 60–62, 73, 76, 78, 81–82, 84, 86–87, 93, 95, 103, 106, 114, 123, 137, 148, 155, 163, 167
Chronicles of Jerahmeel, 43, 139, 149–50
Cicero, 136

Clement of Alexandria: *Excerpts of Theodotus (Excerpta Ex Theodoto)*, 11, 150, 161, 163–64
Cloud of the Impossible, The: Negative Theology and Planetary Entanglement (Keller), 143
co-creation, 8, 41, 60, 63, 101, 111, 130
Cochran, Elizabeth Agnew: "Virtuous Assent and Christian Faith: Retrieving Stoic Virtue Theory for Christian Ethics," 178
Coleridge, Samuel Taylor: *Shakespeare*, 23
Collected Papers vols. I and II: Principles of Philosophy and Elements of Logic (Peirce), 135
collection and division, 64–69, 84, 90, 106
Colossians (Biblical letter), 60, 138, 155–56, 162–63
Colossian Hymn in Context, The: An Exegesis in Light of Jewish and Greco-Roman Hymnic and Epistolary Conventions (Gordley), 155–56
Commentary on Ephesians 4:4 (Jerome), 170
Commentary on Genesis 1:26 (Nachmanides), 169
Commentary on Matthew (Origen), 177
Commentary on Mishnah Sanhedrin (Maimonides), 155
Commentary on Romans (Origen), 177
Commentary on Sefer Yisirah (Isaac the Blind), 152
Commentary on the Gospel According to John (Origen), 59, 171
compassion, 43, 159
"Complete History of Early Christianity, A: Taking the 'Heretics' Seriously" (Franzmann), 139
completeness, 40, 76, 87–88, 90–91
concealment, 48–49, 53–55, 57–58, 74, 86, 151–52. *See also* hiddenness
Conferences, The (John Cassian), 156
"Confession of Faith" (Fourth Lateran Council), 113, 174
Confessions (Augustine), 127, 156, 168, 178
consciousness (human), 3–4, 32
constellations, 69, 100, 140, 176
"Contempts" (Carson), 72
Conway, Kellyanne, 143
Coole, Diana, 2, 4–5, 137; "The Inertia of Matter and the Generativity of Flesh," 136; Introduction to *New Materialisms: Ontology, Agency, and Politics*, 136
Corinthian Body, The (Martin), 96, 167, 172

Corinthians (Biblical letters), 92, 95, 138, 162–63, 165–67, 171–72
Corporal Knowledge: Early Christian Bodies (Glancy), 170
Corpus (Nancy), 92
corruption, 31, 114, 121. 128, 172
cosmogony, 34, 46, 53, 57, 71–72, 74, 76, 80, 93, 141
cosmology, 8, 11–12, 17–18, 47, 63, 71–72, 75, 155, 161
cosmos, 3, 9, 11–12, 18–19, 29, 34, 39, 42–43, 46–48, 50, 52, 56, 58, 69,71, 88, 92–93, 103–4, 108–12, 118–21, 129–30, 154, 173, 176
"Cosmos" (Chin), 9, 138
Cossins, Daniel: "Why We Like to Know Useless Stuff," 141
Cowley, A. E.: *The Original Hebrew of a Portion of Ecclesiasticus*, 147
Cox, Ronald: *By the Same Word: Creation and Salvation in Hellenistic Judaism and Early Christianity*, 137–38
Cratylus (Plato), 145
creation, 3–4, 6–7, 9–11, 18, 20, 33–36, 38–43, 45–47, 51–53, 57, 61, 63–64, 72, 75, 78–80, 85, 88, 93, 99–104, 110–11, 114–15, 119, 121, 123, 126–27, 130, 138, 141, 147, 149, 151–52, 162, 165–66, 168–71, 174
"Creation Theology in the Book of Job" (Bergant), 7, 137
creator, 8, 40–41, 49–50, 55–56, 60–61, 63, 99, 101, 111, 123, 155
creature, 9, 80, 99, 101, 128–29, 168
crucifixion, 61–62
"Cult-Mystery in the Gospel of Philip, A" (Buckley), 171
Curd, Patricia, 19, 29; "Presocratic Philosophy," 142; *A Presocratics Reader*, editor and translator, 141–42, 144–45, 153, 176
cursed, 44, 128, 149

Daniel (Biblical book), 177
darkness, 35, 38, 54–55, 57, 62, 70, 85, 100, 107, 122, 159
Davies, Stevan: *The Secret Book of John: The Gnostic Gospel Annotated and Explained*, editor, translator, and annotator, 49, 52–53, 70, 73, 83, 152–53, 160
death, 40–41, 44, 61–62, 83, 96–101, 103, 108, 121, 128, 130, 142, 148
deficiency, 87–88, 107, 118
deity, 27
delight, 6, 76, 111, 123, 126, 130

Demeter, 62–63
Democracy of Objects, The (Bryant), 2, 136
Democritus, 29, 150
demons, 50, 52, 152, 168
Dennis, Geoffrey W.: "Adam Kadmon," 151
depth, 7, 38, 39, 52, 71, 73, 78, 99
Derech Eretz Zutta, 52, 152
Descartes, René, 21–22, 24, 125; Letter to his translator, 142; *Meditations on First Philosophy*, 22, 125, 135, 142, 178
desire, 23–25, 34–35, 57–58, 73–76, 83, 88, 91, 124–25, 127–28, 131, 165, 178–79
Desire and Delight: A New Reading of Augustine's Confessions (Miles), 127, 179
Deuteronomy (Biblical book), 138, 150, 166
Diagoras of Melos, 29, 144
Dialectic of Enlightenment, The: Philosophical Fragments (Adorno and Horkheimer), 22–24, 37, 143, 146
differentiation, 42–43, 51, 68–69, 74, 161, 178
Dillon, John M.: "Solomon Ibn Gabirol's Doctrine of Intelligible Matter," 70, 159
Diogenes Laërtius: *Lives and Opinions of Eminent Philosophers, Including the Biographies of the Cynics and the Life of Epicurus*, 15, 17, 27, 29, 140, 144, 146
Diotima, 34, 146
Dirac, Paul: "The Quantum Theory of the Electron," 135
dirt. *See* dust
disbelief, 21, 112–13
disciples, 84, 87, 96, 111, 172
disobedience, 44, 55, 61, 102–3
distinction, 2, 8–9, 16–17, 20, 31–32, 35–36, 78, 105–6, 109–10, 122, 124, 127
divinity, 2–13, 19, 21, 23, 27–30, 33–35, 38, 40–41, 43–45, 47–63, 67, 73, 79–83, 87–88, 90–91, 93–94, 99–100, 104–7, 109–11, 116, 118, 120, 123, 125–27, 129–31, 140, 142, 144, 152–53 154–56, 168, 178
Divine Enticement: Theological Seductions (MacKendrick), 136
divinization, 62–63, 89, 154
division, 38, 40–43, 48, 55, 58, 60–62, 74, 76–77, 84, 87, 103, 104, 106–7, 154
Docetism, 156
duality, 26, 40, 65, 71, 97, 105, 140, 151, 155, 157, 159
Dunderberg, Ismo, 11; "Recognizing the Valentinians—Now and Then," 139; "Valentinus and His School," 139, 162, 167

dust, 42–47, 51, 56–58, 93–94, 99–102, 112, 115, 118, 121, 128–30, 134, 149, 169–70. *See also* earth
dyad, 66–71, 71–74, 79, 88, 90–91, 147, 158–59. *See also* Indefinite; Unlimited

Early Church and the Afterlife, The: Post-Death Existence in Athenagoras, Tertullian, Origen, and the Letter to Rheginos (Rankin), 168
earth: as dirt, 6, 41–45, 58, 94, 99, 100–2, 108–9, 111, 115, 123, 128, 149, 172. as planet or realm, 7–8, 12, 26–27, 38–39, 41–45, 47, 52, 58, 60, 62, 94, 100–2, 105, 111, 116, 123–25, 127–28, 149, 151, 162, 166, 176. *See also* dust
Ecce Homo (Nietzsche), 120, 176–77
Ecclesiastes (Bibical book), 7, 137, 174
Ecclesiastes Rabbah (Kohelet Rabbah), 168
Ecclesiasticus. *See* Sirach
Eckhart, Meister: German Sermon on Ecclesiasticus 24:30, 154
Eden, 127, 149. *See also* paradise
educator. *See* teacher
Edwards, Mark J.: "The Epistle to Rheginus: Valentinianism in the Fourth Century," 107, 171–72
Ehrman, Bart D.: "Christianity Turned on Its Head: The Alternative Vision of the Gospel of Judas," 139
ekpyrosis, 119
Elegance of the Hedgehog, The (Barbery), 1
Elementa Harmonica, 158
Eleusinian mysteries, 62, 144
Elijah (Biblical figure), 104
Eliot, T. S.: "Burnt Norton," 45, 172
Elson, Rebecca: *A Responsibility to Awe*, 20, 75
emanation, 41, 47–48, 53, 64, 69–70, 164, 170
embodiment, 12, 35, 40, 50, 61–62, 77, 79–83, 93, 98, 103–4, 109–13, 121, 124, 128, 172, 174. *See also* body; incarnation
embryo, 46, 150
Emerald Tablet, The, 150
emotion, 32, 35–36; creating matter, 76–77. *See also* passion
Empedocles, 18, 112, 120, 165, 176
"Empedocles of Agrigentum" (Kohler and Broydé), 165
Enchantment of Modern Life, The: Attachments, Crossings, and Ethics (Bennett), 143
Enlightenment, 2–24, 37, 142–43

Enneads, The (Plotinus), 153, 158
Enoch. *See* 1 Enoch; 2 Enoch
Entangled Worlds: Religion, Science, and New Materialisms (Keller and Rubenstein, editors), 135
Entireties, 51, 77, 153. *See also* Aeons; All, The; Fullness; Pleroma
Ephesians, 103, 162–64, 170
Epicurus: *The Essential Epicurus: Letters, Principal Doctrines, Vatican Sayings, and Fragments*, 23, 140, 143
Epinoia, 53–54, 155
epistemology, 3–4, 50, 52, 56, 64, 66, 89–91, 135, 142
Epistle of John (Biblical letter), 61
Epistle to Rheginus (also *Treatise on the Resurrection*), 96, 105–7, 124, 156, 167–68, 171–72
"Epistle to Rheginus, The: Valentinianism in the Fourth Century" (Edwards), 107, 171–72
Eros, 18, 24, 34, 64, 90–91
Eros the Bittersweet (Carson), 161
eschatology, realized, 96–98, 102, 105–7, 110, 118–19, 122, 124
Essays on the Book of Enoch and Other Early Jewish Texts and Traditions (Knibb), 155
Essenes, 94, 166
Essential Epicurus, The: Letters, Principal Doctrines, Vatican Sayings, and Fragments (Epicurus), 23, 140, 143
eternal recurrence. *See* recurrence
Eve (Biblical and mythical figure), 39–40, 53, 55, 147, 155
"Everything is Water" (Rojcewicz), 20, 142
ex nihilo, 99, 115
Excerpta Ex Theodoto of Clement of Alexandria (Clement of Alexandria), 11, 150, 161, 163–64
Exodus (Biblical book), 149, 151, 173
expulsion from paradise, 4, 44, 55, 102, 128, 149
eye that sees itself, 52–53, 56–57, 122, 154
Ezekiel (Biblical figure), 60, 94, 100, 155, 166, 169, 172

fact, 24–26, 31–37, 124–25, 127, 130, 143
faith, 8, 10, 21, 25, 41, 59, 62, 95–96, 105, 127, 153, 165, 178–79
falsehood, 26, 31, 34. *See also* fiction; myth
Father, The, 51–52, 60–62, 64, 71–80, 82–83, 86–88, 90–91, 106, 123, 152, 159–61, 164–65

Fathers According to Rabbi Nathan, The, 47, 151
"fehlehnden Bände des Römerbriefkommentars des Origenes, Die" (Bammel), 178
feminine, 70–71, 130
fern, Barnsley, 88
fiction, 31–32, 67, 140. *See also* falsehood; myth
Filoramo, Giovanni: *A History of Gnosticism*, 139
Fine, Lawrence: "Tikkun: A Lurianic Motif in Contemporary Jewish Thought," 154, 174
Firmicus Maternus, 150
Flavius Josephus: *The Genuine Works of Flavius Josephus the Jewish Historian*, 93–94, 96, 166
flesh: embodying divinity, 8, 45, 58, 60–61, 72, 83–86, 90–92, 129, 150, 163, 168; fluid, 4–5, 64, 70, 79, 150; and forgetfulness, 62–63, 80–81, 98, 102; and form, 27, 30, 38–39, 41, 43, 45–46, 49–54, 56, 60, 65–66, 69–70, 74, 76–80, 88–89, 102, 109–10, 112–15, 118, 122, 128–29, 152, 170–71, 174, 177; formlessness of, 38–39, 41, 43, 46, 50–52, 68, 70, 73, 80, 99, 152; of human body, 6, 12, 39–40, 42, 44–45, 50, 58, 62, 85, 89, 93–100, 102, 104–7, 109–10, 114–18, 124, 129, 150, 166, 168–69, 171, 174–75; of world, 12–13, 57, 60, 85, 89–92, 109–10, 129, 172
Foucault, Michel, 35–37; *The Hermeneutics of the Subject: Lectures at the Collège de France 1981–1982*, 146; "Nietzsche, Genealogy, History," 63
Four Fundamental Concepts of Psychoanalysis (Lacan), 154
Fowler, Robert: "Mythos and Logos," 31, 145
fractals, 48, 88–91, 106, 164
"Fractal Geometry Is Not the Geometry of Nature" (Shenker), 89, 164
Fractal Geometry of Nature, The (Mandelbrot), 89, 164
"Fractals, Wavelets, and Their Applications" (Bandt), 66, 164
Fragments: A Text and Translation With a Commentary (Xenophanes), 30, 142, 145
Franke, William: "Poetry, Prophecy, and Theological Revelation," 146
Franzmann, Majella: "A Complete History of Early Christianity: Taking the 'Heretics' Seriously," 139

INDEX

freedom, 5, 23, 81, 90, 126
Freud, Sigmund 127; *Instincts and their Vicissitudes*, 179; *Three Essays on the Theory of Sexuality*, 179
"From Periphery to Center: Early Discussions of Resurrection in Medieval Jewish Thought" (Schlossberg and D. Schwartz), 168, 175
"From What Is One and Simple Only What Is One and Simple Can Come to Be" (Hyman), 159
Frost, Samantha: Introduction to *New Materialisms: Ontology, Agency, Politics*, 4–5, 136–37
Fullness, 77–79, 161. *See also* Aeons; All, The; Pleroma
futurity (of resurrection), 96–98

Gabriel (angel), 43–44
Gaiser, Konrad: "Plato's Enigmatic Lecture 'On the Good,'" 67, 157–58
Garber, Marjorie: *The Muses on Their Lunch Hour*, 25, 143
Gardet, L.: "Kiyāma," 175
garden, 128. *See also* Eden; paradise
Gardens: An Essay on the Human Condition (Harrison), 127, 179
garment, 51, 55, 111, 153–54
Gast, Peter, 120, 176
Gate of Resurrections, The (Luria), 111, 173–74
Gay Science, The (Nietzsche), 146, 176–77
gematria, 151
gender, 41, 43, 71. *See also* androgyny; feminine
Genesis (Biblical book), 3–4, 6–8, 12, 26, 38–39, 42, 46, 53–55, 58–61, 63, 85, 100, 102, 110–11, 127, 136–37, 139, 147, 151–52, 155–56, 160, 164, 173, 177
Genesis Rabbah, 40, 43, 48–50, 55, 92, 101–2, 153–54, 163, 165, 168–70
geometry, 65, 70, 89, 164
German Sermon on Ecclesiasticus 24:30 (Eckhart), 154
Gilbert, Maurice: "*Pirqé Avot* and Wisdom Tradition," 138
Gilgamesh, 7
gilgul. *See* reincarnation
Glancy, Jennifer: *Corporeal Knowledge: Early Christian Bodies*, 170
glorious body, 11–12, 92, 121, 123. *See also* resurrection
gnosis, 10, 62, 81–82, 106, 122, 126, 153, 161

Gnosis: An Introduction (Markschies), 162
Gnosticism, 10–11, 40–41, 46, 50, 54, 56, 63, 70–71, 73, 80–81, 88, 90–91, 121, 134, 139, 156, 161
"Gnosticism" (Moore), 164–65
Gnostic Paul, The: Gnostic Exegesis of the Pauline Letters (Pagels), 172
Gnostic Religion, The: The Message of the Alien God and the Beginnings of Christianity (Jonas), 139
Gnostics, The: Myth, Ritual, and Diversity in Early Christianity (Brakke), 139
Goff, Matthew: "Searching for Wisdom in and Beyond 4*Qinstruction*," 137
Goodman, Lenn E.: "Ibn Masarrah," 177; Introduction to *Neoplatonism and Jewish Thought*, 67, 158
Gordley, Matthew: *The Colossian Hymn in Context: An Exegesis in Light of Jewish and Greco-Roman Hymnic and Epistolary Conventions*, 155–56
Gospel of John, The, 8, 60–61, 75, 97, 138
Gospel of Philip, The, 40, 42, 55, 57, 81, 86, 102–3, 105–7, 123, 126, 148–49, 153–54, 164, 170–72, 178–79
Gospel of Thomas, The, 80, 96, 102, 170, 172
Gospel of Truth, The, 62, 72–73, 76, 82–83, 86–87, 107, 126, 156, 160, 162–64, 179
gospels, 93, 95. *See also specific titles*
Gregory, Andrew: *The Presocratics and the Supernatural*, 27
Gregory of Nyssa, 103, 113–14, 122; "On the Soul and Resurrection," 113, 170, 174, 177
Grill, Julias, 160
Guthrie, W. K. C.: *A History of Greek Philosophy*. Vol. 1, *The Earlier Presocratics and the Pythagoreans*, 141; *History of Greek Philosophy*, A. Vol. 5, *Later Plato and the Academy*, 67, 158

Halevi, Judah: *Sefer Kuzari*, 148
Harmon, Graham: "On Vicarious Causation," 137; "The Road to Objects," 3, 136–37
harmony, 35, 68, 77, 86, 103, 158
Harrison, Robert Pogue: *Gardens: An Essay on the Human Condition*, 127, 179
Hazard, Sonia: "The Material Turn in the Study of Religion," 136
heaven, 28, 44, 46, 51, 60, 95–96, 103, 109, 111, 119, 130, 151, 162, 167, 172, 176
Hebrews (Biblical letter), 138, 155
Hegel, G. W. F., 127–28

Heidegger, Martin: *Introduction to Metaphysics*, 3, 135
Hendy, Andrew von: *The Modern Construction of Myth*, 145
Heraclitus, 18, 54, 93, 146, 153, 176
heresiology, 10–11, 27, 72, 90, 96–97, 107. *See also specific authors and titles*
hermaphroditism. *See* androgyny
Hermeneutics of the Subject, The: Lectures at the Collège de France 1981–1982 (Foucault), 146
hermeticism, 150
"Here Comes Everything: The Promise of Object-Oriented Ontology" (Morton), 136–37
Hesiod: *Theogony*, 17–18, 27, 29, 141
hiddenness, 44, 47–49, 57, 62, 86, 90, 109, 122, 159. *See also* concealment
hierarchy, 64–65, 170
Hippo of Samos, 29, 144
Hippolytus: *Refutation of All Heresies*, 27–28, 70–71, 77, 91, 144, 159, 161–62, 165
History of Gnosticism, A (Filoramo), 139
Holland, Frederic May: *The Rise of Intellectual Liberty from Thales to Copernicus*, 142
hologram, 85, 87–88, 90–91, 106
Homer: *The Iliad*, 141; *The Odyssey*, 18, 27, 29, 142
Horkheimer, Max: *The Dialectic of Enlightenment: Philosophical Fragments*, 22–24, 37, 143, 146
Horos, 73, 76, 84, 160–61, 163. *See also* Limit
Horst, Pieter W. van der: "The Measurement of the Body: A Chapter in the History of Ancient Jewish Mysticism," 150
"How to be an Atheist" (D. Turner), 30, 145
Huffer, Lynne: *Mad for Foucault: Rethinking the Foundations of Queer Theory*, 34, 146
human, 2–6, 9, 11, 22–23, 25–26, 39–45, 47, 49, 52, 54–55, 62, 90, 95, 102, 110–11, 114–17, 127–31, 142, 156, 169; as image of divinity, 3, 6, 39–42, 44–45, 47–55, 57–58, 60, 118, 123, 152
humanities, 1, 3, 25, 136, 143
Hyman, Arthur: "From What Is One and Simple Only What Is One and Simple Can Come to Be," 159

Iamblichus, 137, 145–46, 173
ibn Ezra, Abraham, 41, 46–47, 170; *The Book of Ibn Ezra*, 148, 150–51
ibn Gabirol, Solomon, 66, 159, 165
Ibn Gabirol's Theology of Desire: Matter and Method in Jewish Medieval Neoplatonism (Pessin), 157, 165
ignorance, 31, 74, 76, 82
Iliad, The (Homer), 141
illimitable, 49, 58, 72, 80
illusion, 2, 16, 61, 105, 118
imagination, 2, 9, 31–32, 64, 93, 152
immanence, 1–2, 5–6, 62, 96–98, 105, 118, 121, 124, 130
immeasurable, 13, 49, 51
immortality, 6, 21, 95, 110, 119, 124, 142, 166, 173. *See also* resurrection
Immortality, Resurrection, and the Age of the Universe: A Kabbalistic View (Kaplan), 165
imperfection, bodily, 50, 80, 114, 117
imperishable flesh, 92, 95, 107, 172
In the Eye of the Animal: Zoological Imagination in Ancient Christianity (P.C. Miller), 136
incarnation, 12, 60, 81, 83, 107–10, 119, 168, 176. *See also* body; embodiment
incarnation, successive, 103, 108–12, 118–19, 122, 124
incomprehensibility, 5, 51, 57, 83, 86
incorruption, 121, 152, 172
Indefinite, 66–70, 158. *See also* dyad; Unlimited
ineffability, 33, 35, 72 74, 80, 153. *See also* unknowable
"Inertia of Matter and the Generativity of Flesh, The" (Coole), 136
Infinite Conversation, The (Blanchot), 146
infinite, 28, 38, 48–49, 68, 82, 88, 90, 93, 106, 119–20, 124, 146, 164
Inhofe, James, Senator, 26, 143
Inner Grace: Augustine in the Traditions of Plato and Paul (Cary), 178
Instincts and their Vicissitudes (Freud), 179
Intercarnations: Exercises in Theological Possibility (Keller), 135
interchange of matter, 9, 113–20
Interpretation of Knowledge, The, 81
Intoxication (Nancy), 31
"Introduction to Fractals" (Bandt), 88, 164
Introduction to Metaphysics (Heidegger), 3, 135
Introduction to *Religious Experience and New Materialism: Movement Matters* (Jones), 1, 135
Introduction to the Reading of Hegel: Lectures on the Phenomenology of Spirit (Kojève), 179

INDEX

Irenaeus: *Against Heresies*, 11, 45, 63, 71, 73, 75–76, 78, 84, 87, 90–91, 95–97, 99–100, 107, 126, 156, 159–65, 167–69
Isaac the Blind: *Commentary on the Sefer Yisirah*, 152
Isaiah (Biblical book), 94, 100, 166, 169
Islam, 9, 21, 159, 165, 168, 175
Israel, 59, 101, 137–38, 153–54, 171, 174

Jerome: His Life, Writings, and Controversies (Kelly), 178
John (Biblical book). *See* The Gospel of John
John, Epistle of (Biblical letter), 61
John Cassian, *The Conferences*, 156
Jonah (Biblical figure), 175
Jonas, Hans: *The Gnostic Religion: The Message of the Alien God and the Beginnings of Christianity*, 175
Jones, Tamsin: Introduction to *Religious Experience and New Materialism: Movement Matters*, 1, 135
joy, 24, 76–77, 80, 94, 98, 101, 106, 120, 126–27, 128, 176. *See also* delight
Judaism, 10, 59, 137–39, 154, 174. *See also* Kabbalah; rabbinic Judaism
Judges (Biblical book), 177
judgment, 43, 65, 97–98, 101, 106, 112–13, 125, 130, 170–71

Kabbalah, 11, 40–41, 46–47, 51, 54, 57, 70, 92, 100, 104, 106, 110, 147, 151, 154–55, 159, 165, 173–74. *See also specific titles and authors*
Kalonymus Kalman Shapira, Rabbi, 165
Kalvasmaki, Joel: *The Theology of Arithmetic: Number Symbolism in Platonism and Early Christianity*, 69, 71, 73, 89, 158–60, 164
Kant, Immanuel: "An Answer to the Question: What Is Enlightenment?," 22, 142
Kaplan, Aryeh: *Immortality, Resurrection, and the Age of the Universe: A Kabbalistic View*, 165
Keller, Catherine, 2; *Cloud of the Impossible: Negative Theology and Planetary Entanglement*, 143; *Entangled Worlds: Religion, Science, and New Materialisms*, 135; *Intercarnations: Exercises in Theological Possibility*, 135
Kelly, J. N. D.: *Jerome: His Life, Writings, and Controversies*, 178
Kenny, John P.: "The Platonism of the Tripartite Tractate (Nh I.5)," 77, 88, 162, 164
Kiddushin, 155

King, Karen: *The Secret Revelation of John*, editor, translator, and annotator, 117
kingdom of God, 96, 102, 105, 172
Klossowski, Pierre: "Nietzsche's Experience of the Eternal Return," 121, 177
Knibb, Michael Anthony: *Essays on the Book of Enoch and Other Early Jewish Texts and Traditions*, 155
knowing, 2, 3, 6, 9–10, 12, 16, 18–19, 22–23, 25, 30, 32, 44, 52, 54–57, 64–5, 90, 109, 128, 142–43, 161; as gnosis or redemption, 63, 81–83, 85–87, 91, 106–7, 110, 123–24, 126, 129; limits of, 1, 3–5, 31, 33–37, 49, 50–51, 57–58, 65, 72, 74–75, 82–83, 85, 91, 118; and love or desire, 65, 74–76, 90–92, 123, 126–27, 129, 161, 165
Koch snowflake, 88, 164
Kohelet Rabbah. *See Ecclesiastes Rabbah*
Kohler, Kaufmann: "Empedocles of Agrigentum," 165
Kojève, Alexandre: *Introduction to the Reading of Hegel: Lectures on the Phenomenology of Spirit*, 179
"Kotnot Or (Genesis 3:21): Skin, Leather, Light, or Blind?" (Schneider and Seelenfreund), 154
Kramer, Hans Joachim: *Plato and the Foundations of Metaphysics: A Work on the Theory of the Principles and Unwritten Doctrines of Plato with a Collection of the Fundamental Documents*, 157

Lacan, Jacques: *The Seminar of Jacques Lacan: Four Fundamental Concepts of Psychoanalysis*, 154; *The Seminar of Jacques Lacan: The Ego in Freud's Theory and in the Techniques of Psychoanalysis, 1954–55*, 179
Laitman, Michael: *Unlocking the Zohar*, 173
Lambert, Wilfred G.: *Babylonian Wisdom Literature*, 138
Language, Eros, Being (Wolfson), 147, 173
Latour, Bruno: "On Actor-Network Theory: A Few Clarifications," 6, 137; "On Recalling ANT," 137; *On the Modern Cult of the Factish Gods*, 25, 143; *Pandora's Hope: Essays on the Reality of Science Studies*, 137
Layton, Bentley, 171
Lebesgue-Osgood monsters, 89
Leviticus (Biblical book), 40, 54, 150, 153, 165, 168

Leviticus Rabbah (Vayikra Rabbah), 148, 166
"Life Full of Meaning and Purpose, A: Demiurgical Myths and Social Implications" (Williams), 139
light, 38–39, 47, 49, 52–58, 61, 76, 85, 93, 100, 105–7, 111, 122–23, 153–54, 159, 177. *See also* brightness; luminosity
likeness, 25, 39, 40, 42, 44, 46–47, 52, 77, 111, 151, 160
Lilith, 147
limbs, 50–51, 112, 149, 152
limit, 8, 37–39, 43, 46, 51, 57–58, 63, 66–70, 72–78, 80, 82–83, 85, 88–91, 93, 160, 162, 164. *See also* boundary
Limit, 63–64, 66–69, 72–85, 88, 106, 130, 161, 163, 165. *See also* Horos
Lives and Opinions of Eminent Philosophers, Including the Biographies of the Cynics and the Life of Epicurus (Diogenes Laërtius), 15, 17, 27, 29, 140, 144, 146
logic, 11, 19, 26, 31–32, 34, 37, 125, 135
logos, 18, 31–32, 145, 156
Logos, 8, 60, 70
Lore and Science in Ancient Pythagoreanism (Burkert), 158
love, 15, 24, 34, 61, 65, 75, 76, 81, 88, 90–91, 123, 125–28, 141, 153–54, 164–65, 177–79
Luke (Biblical book), 139, 163, 166, 168, 174
luminosity, 18, 24, 52–56, 81, 110, 122. *See also* brightness; light
Lupieri, Edmondo: *The Mandaeans: The Last Gnostics*, 139
Luria, Isaac, Rabbi, 47, 54–55, 92, 110–11, 113, 118, 154, 173–74; *Collected Works on Kabbalah*, 154; *Gate of Reincarnations*, 173–74. *See also* Vital, Rabbi Chaim
luz bone, 100, 112

Maccabees (Biblical book), 166, 169
Macfarlane, Robert: *Underland*, 43, 107
macrocosm, 48. *See also* microcosm
Mad for Foucault: Rethinking the Foundations of Queer Theory (Huffer), 34, 146
Maimonides, Moses, Rabbi (Moses ben Maimon), 59–60, 97; "The 13 Principles and the Resurrection of the Dead," 167; *Commentary on Mishnah Sanhedrin*, 155; *Mishnah Torah*, 155, 168; *Treatise on Resurrection*, 168
Malter, Henry: *Saadia Gaon: His Life and Works*, 168, 177

Man and Temple: In Ancient Jewish Myth and Ritual (Patai), 150–51
Mandaeans, The: The Last Gnostics (Lupieri), 139
Mandelbrot, Benoit: *The Fractal Geometry of Nature*, 89, 164
Manichaeans, 140, 156, 173
Mansfeld, Jaap: "Aristotle and Others on Thales, or the Beginnings of Natural Philosophy," 141
Marcus (Valentinian teacher), 87, 90
Markschies, Christoph: *Gnosis: An Introduction*, 162
Martin, Dale: *The Corinthian Body*, 96, 167, 172
martyr, 115, 123
"Material Turn in the Study of Religion, The" (Hazard), 136
materialism, 1–7, 5, 9, 11, 17, 20–21, 25, 33, 44, 52, 57, 62–64, 70, 76, 83, 91, 102, 118, 130, 135–37, 152, 161
mathematics, 65, 67, 88–89, 135, 139. *See also* arithmetic; number
Mathesis, 150
Matt, Daniel: *Zohar*, translator, editor, and annotator, 173, 148, 159
matter: inert or agential, 2–6, 9, 24–25, 43, 45, 50, 69, 91, 118, 124, 127–29, 179; regathered, 101–2, 104, 110–17; transformed, 38, 45, 63, 96–97, 101–2, 107–8, 118, 122, 124–27
Matthew (Biblical book), 163, 166, 174, 177
McGlothlin, Thomas Dwight: *Raised to Newness of Life: Resurrection and Moral Transformation in Second- and Third-Century Christian Theology*, 177
"Measurement of the Body, The: A Chapter in the History of Ancient Jewish Mysticism" (van der Horst), 150
Medieval thought, 6, 11, 21, 30, 40–41, 66, 70, 142, 154, 157, 159, 167–68. *See also specific writers and texts*
Meditations on First Philosophy (Descartes), 21–22, 135, 142, 178
"Meditations in an Emergency" (O'Hara), 130, 179
mending, 104, 110, 112. *See also* tikkun
Merleau-Ponty, Maurice: *The Visible and the Invisible*, 57, 129, 154, 179
messiah, 59–60, 62, 96, 103–4, 155
Metaphysics (Aristotle), 66, 140–41, 143–45, 157–59, 179
metempsychosis. *See* reincarnation

INDEX   209

Meyer, Marvin: "Gnosticism, Gnostics, and the Gnostic Bible," 56, 154; *Nag Hammadi Scriptures*, editor, 148, 152, 154, 156, 160, 162, 165, 170–71, 179
microcosm, 45–52, 56, 93, 103, 110, 152, 154
middle Platonism. *See* Platonism
Miles, Margaret: *Desire and Delight: A New Reading of Augustine's Confessions*, 127, 179
"Milesian Measures: Time, Space, and Matter" (White), 142
Milesian monism, 16–18, 22, 24, 28, 142, 144, 176. *See also specific names*
Miller, Mitchell: "The Choice between the Dialogues and the 'Unwritten Teachings': A Scylla and Charybdis for the Interpreter?," 157
Miller, Patricia Cox, 86, 90; "Adam, Eve, and the Elephants: Asceticism and Animality in Late Ancient Christianity," 136; *In the Eye of the Animal: Zoological Imagination in Ancient Christianity*, 136; "'Plenty Sleeps There': The Myth of Eros and Psyche in Plotinus and Gnosticism," 156, 164–65; "'Words with an Alien Voice': Gnostics, Scripture, and Canon," 164
Mishnah Gittin, 171
Mishnah Ketubot, 171
Mishnah Sanhedrin, 104, 149, 155, 165–68
Mishnah Torah (Maimonides), 155, 168
Mnemosyne, 141
Moderatus of Gades, 69–70, 89
Modern Construction of Myth, The (von Hendy), 145
Moffatt, James: "Pistis Sophia," 160
Moll, Sebastian: *The Arch-Heretic Marcion*, 139
monad, 71, 159. *See also* One, The
monism, 16, 18, 28 71, 159. *See also* Milesian monism
Moore, Edward: "Gnosticism," 164–65
Morton, Timothy: "Here Comes Everything: The Promise of Object-Oriented Ontology," 136–37
Moses (Biblical figure), 101, 175
multiplicity, 41–42, 49, 65, 68–70, 74, 77, 79, 86, 88, 90–91, 111, 113, 121, 155, 171, 174
muses, 18–19, 141
Muses on their Lunch Hour, The (Garber), 25, 143
Mystagogie de Proclos, Le (Trouillard), 146
mysteries (cultic), 23, 44, 62, 67, 100, 129, 137, 144

mystery, 1–2, 4–5, 7, 10, 12, 24, 30, 33–34, 36–37, 43, 45, 47–48, 54, 56–58, 62, 67, 69, 71, 91–92, 98, 104, 106–7, 110, 120, 129–31, 147, 153, 157, 171–72. *See also* mysticism
"Mystical Rationalization of the Commandments in *Sefer ha-Rimmon*" (Wolfson), 151
mysticism, 2, 7, 56, 130, 151–52
myth, 2–3, 9, 12–13, 15–29, 31–37, 45, 57, 64, 70, 84, 126, 129–30, 139–40, 143, 145–46, 145, 151, 156, 161
"Myth and Scripture: Dissonance and Convergence" (Callender), 145
"Myth, History, and Dialectic in Plato's Republic and Timaeus-Critia" (Rowe), 36, 146
"Mythos and Logos" (Fowler), 31, 145
"Mythos and Logos" (Orfanos), 32, 145

Nachman of Bratslav, Rabbi, 165
Nachmanides (Moses bin Nachman), 41–42, 101, 109–11, 118, 148, 169, 171, 173; *Sh'ar ha-Gemul*, 173
"Nachmanides' Commentary on the Book of Job" (Silver), 94, 166, 173
Nag Hammadi texts, 138, 140, 148, 152, 154, 156, 159–60, 162, 165, 167, 170–71, 173, 179. *See also specific titles*
name, 47, 64, 83–88, 90, 151, 164, 171
namelessness, 57, 90
Nancy, Jean-Luc: 31, 37, 92, 146; *Corpus*, 92; *Intoxication*, 31; *The Pleasure in Drawing*, 121
"Nature's Queer Performativity" (Barad), 5, 137
Nemesius of Emesa, Bishop, 119, 176
Neo-Platonism (also Neoplatonism), 4, 33, 35, 53, 68–70, 125, 137, 156–59, 162, 172. *See also specific authors, specific titles*
Neo-Pythagoreanism, 69, 72, 89, 165
Neubauer, A.: *The Original Hebrew of a Portion of Ecclesiasticus*, 147
New Materialisms: Ontology, Agency, Politics (Coole and Froste, editors), 136–37
Nietzsche, Friedrich, 35, 63, 119–21, 124, 146, 176–77; *Ecce Homo: How One Becomes What One Is*, 120, 176–77; *The Gay Science*, 146, 176–77; *Philosophy in the Tragic Age of the Greeks*, 176; *The Selected Letters of Friedrich Nietzsche*, 176; *Thus Spoke Zarathustra*, 176; *Untimely Meditations*, 176; *The Will to Power*, 176

"Nietzsche's Experience of the Eternal Return" (Klossowski), 121, 177
"Nietzsche, Genealogy, History" (Foucault), 63
Noah (Biblical figure), 55
"Note on the Rabbinic Doctrine of Creation, A" (Altmann), 170
number, 35, 71, 113–14, 118–19, 174. *See also* arithmetic; mathematics

object oriented ontology, 2–3, 5, 136
Oceanus, 18, 141
O'Dowd, Ryan P.: *Old Testament Wisdom Literature: A Theological Introduction*, 138
Odyssey, The (Homer), 18, 27, 29, 142
Of Silence and Sound (Beachy-Quick), 38
O'Grady, Patricia: *Thales of Miletus: The Beginnings of Western Science and Philosophy*, 27–28, 144
O'Hara, Frank, "Meditations in an Emergency," 130, 179
Old Testament Wisdom Literature: A Theological Introduction (Bartholomew and O'Dowd), 138
Olsen, Roger E.: *The Story of Christian Theology: Twenty Centuries of Tradition and Reform*, 140
"On Actor-Network Theory: A Few Clarifications" (Latour), 6, 137
On First Principles (Origen), 162, 170–71, 173, 177
On First Principles: Known as His Metaphysics (Theophrastus), 19, 158
On Modesty (Tertullian), 162
On Nature (Peri Physeos) (Parmenides), 140
On Prayer (Origen), 172
"On Recalling ANT" (Latour), 137
On the Modern Cult of the Factish Gods (Latour), 25, 143
On the Origin of the World, 165
On the Resurrection (Athenagoras), 168, 174–75
On the Resurrection (Tertullian), 150, 175
On the Soul (De Anima) (Aristotle), 114, 144
On the Soul and Resurrection (Gregory of Nyssa), 113, 170, 174, 177
On the Trinity (Augustine of Hippo), 148, 167
"On Vicarious Causation" (Harman), 137
One, The, 53, 65–74, 79, 88, 91, 157, 160. *See also* monad

Orfanos, Spyros: "Mythos and Logos," 32, 145
Origen and the Life of the Stars: A History of an Idea (Scott), 138, 172
Origen of Alexandria, 11, 59, 80, 103–4, 108–11, 113, 118–19, 122, 138, 168, 173; *Against Celsus*, 172, 176; *Commentary on Matthew*, 177; *Commentary on Romans* 5.8.13, 177; *Commentary on the Gospel According to John*, 171; *On First Principles*, 162, 170–71, 173, 177; *On Prayer*, 172
origin, 15–18, 20, 27–28, 29–30, 40, 42, 51, 54, 57–58, 60, 69, 71–72, 74–75, 78–79, 89, 109, 119, 124, 147, 169
Original Hebrew of a Portion of Ecclesiasticus, The (Cowley and Neubauer), 147
orthodoxy, 10, 44, 61, 71, 95–96, 98, 105–6, 112, 139, 148, 154, 156

Pagels, Elaine: *The Gnostic Paul: Gnostic Exegesis of the Pauline Letters*, 139, 172
Palace of Pearls, A: The Stories of Rabbi Nachman of Bratslav (H. Schwartz), 156, 165
Pandora's Hope: Essays on the Reality of Science Studies (Latour), 137
panentheism, 29
pantheism, 29, 144
Pantheologies: Gods, Worlds, Monsters (Rubenstein), 143–44
paradise, 4, 41, 44, 46, 54–55, 127–28. *See also* Eden
paradox, 45, 58, 61, 69, 71, 74, 86, 88, 147
Parmenides: *On Nature (Peri Physeos)*, 16, 18, 140–41
Parmenides, The (Plato), 65–67, 157–58
Partenie, Catalin, 33, 36, 145–46
passion, 31, 57, 76, 78–79, 152, 178
Patai, Raphael: *Man and Temple: In Ancient Jewish Myth and Ritual*, 150–51
Paul (Biblical author), 60–61, 81, 92, 94–96, 103–5, 107–9, 114, 117, 121, 138, 162, 166–67, 172. *See also specific titles*
Pearson, Birger: *Ancient Gnosticism: Traditions and Literature*, 159
Peel, Malcolm, 171
Peirce, Charles Sanders; *Collected Papers of Charles Sanders Peirce. Vol. 1, Principles of Philosophy* and *Collected Papers of Charles Sanders Peirce. Vol. 2, Elements of Logic*, 135
Pentateuch with Rashi's Commentary, 148, 169

Pepple, John, "The Unwritten Doctrines: Plato's Answer to Speusippus," 158
perfection, 15, 32, 40, 49, 57, 69, 93, 97–98, 103–4, 108–11, 114, 118, 122, 155, 173, 175
Peri Physeos. See *On Nature* (Parmenides)
perishing, 28, 95, 99, 101, 107, 171–72, 175
peros, 67. *See also* limit
Pessin, Sarah: *Ibn Gabirol's Theology of Desire: Matter and Method in Jewish Medieval Neoplatonism*, 157, 165
"Peter Quince at the Clavier" (Stevens), 119
Phaedo (Plato), 142, 145
Phaedrus (Plato), 64, 157, 166, 179
Pharisees, 93–95, 166
Philebus (Plato), 66–68, 74, 158
Philo of Alexandria, 4, 38, 70, 147
"Philosopher and the Volcano, The" (Babich), 176
philosophy, 2, 4, 7–9, 11–13, 15–22, 25, 27–33, 36, 57, 61, 64–65, 67, 119–20, 135, 139, 141–46, 150, 158, 164–65, 172, 176, 178–79
Philosophy in the Tragic Age of the Greeks (Nietzsche), 176
physics, 18–19, 25, 64, 143
Physics (Aristotle), 28, 66, 141, 144, 150
Pirkei Avot (also *Pirqé Avot*), 138–39
"Pirqé Avot and Wisdom Tradition" (Gilbert), 138
"Pistis Sophia" (Moffatt), 160
Plato: *Apology*, 4, 136, 142, 163; *Cratylus*, 145; *Parmenides*, 65–67, 157–58; *Phaedo*, 142, 145; *Phaedrus*, 64, 157, 166, 179; *Philebus*, 66–68, 74, 158; *Protagoras*, 18, 32; *Republic*, 145, 153, 157; *Sophist*, 31–32, 145; *Statesman*, 65, 67, 157; *Symposium*, 30, 33–34, 40, 65, 145–46, 148, 157, 179; *Theaetetus*, 18, 65, 67, 141, 157; *Timaeus*, 145–46, 150, 158. *See also* Platonism
Plato and the Foundations of Metaphysics: A Work on the Theory of the Principles and Unwritten Doctrines of Plato with a Collection of the Fundamental Documents (Kramer), 157
Platonism, 68–70, 98, 125–27, 138–40, 155, 159, 165. *See also* Neo-Platonism
"Platonism of the Tripartite Tractate, The" (Kenny), 162, 164
"Plato's Enigmatic Lecture 'On the Good'" (Gaiser), 67, 157–58
Plato's Late Ontology: A Riddle Resolved (Sayre), 65–67, 157–58
"Plato's Myths" (Partenie), 145–46

Plato's Myths (Partenie), 145
"'Plenty Sleeps There': The Myth of Eros and Psyche in Plotinus and Gnosticism" (P.C. Miller), 156, 164–65
Pleroma, 49, 51, 72, 75–77, 84, 88, 107, 156. *See also* Aeons; All, The; Fullness
Plotinus; *Enneads*, 153, 158
"Poem" (Rukeyser), 129
poetry, 11–12, 16, 18–19, 23, 25–26, 29–33, 35, 37, 39, 64, 119, 129–30
"Poetry, Prophecy, and Theological Revelation" (Franke), 146
Popper, Karl: 1958 Presidential Address to the Aristotelian Society, 15
Porphyry, 158; *The Life of Pythagoras*, 108, 172, 176
possibility, 7, 11, 23, 43–44, 60, 62, 72, 79, 96, 99–102, 108, 116, 119, 126, 129–31, 148–49, 156, 176
Posthumanities: Universe of Things; On Speculative Realism (Shaviro), 137
preexistence, 70, 78–79, 99, 115
Prescription Against Heretics (Tertullian), 163
Pre-Socratic, 15–16, 18–19, 120, 142, 144. *See also specific names, specific titles*
"Presocratic Cosmologies" (Wright), 18, 141
Presocratics and the Supernatural, The (A. Gregory), 27
Presocratics Reader, A (Curd, editor, translator, and annotator), 141–42, 144–45, 153, 176
primal human, 11–12, 38, 46–47, 49, 57, 61–62, 67, 71, 75, 90, 92, 104, 118, 147
principle, 27–29, 34, 61, 64, 66, 72, 141, 156, 160. *See also* Logos
Process and Reality (Whitehead), 15, 140
Pronoia, 155
Protagoras (Plato), 18, 32
Proverbs (Biblical book), 7–8, 137
Psalms (Biblical book), 7, 36, 94, 104, 153, 155, 166, 169
psyche, 27, 44, 50, 70, 125, 156
Pythagoras, 15, 35, 64, 69, 145–46, 150, 158, 172, 176
Pythagoras: His Life, Teaching, and Influence (Riedweg), 158
Pythagorean, 33, 35, 46, 64–67, 69–71, 93, 108, 119, 126, 146, 150, 157, 159–60, 173, 176. *See also* Neo-Pythagorean

quantity, 24, 36, 70, 120, 176
"Quantum Theory of the Electron, The" (Dirac), 135

Qumran, 7, 10, 94, 137, 155
Qur'an, 4

rabbinic Judaism, 10–12, 40, 43–48, 54, 59, 80, 93–94, 104, 106, 110–11, 118, 138–39, 151, 153, 165, 170
radiance, 54, 58, 92, 122, 127, 130. See also luminosity
Raised to Newness of Life: Resurrection and Moral Transformation in Second- and Third-Century Christian Theology (McGlothlin), 177
Rambam. See Maimonides, Moses
Ramban. See Nachmanides
Rankin, David: *The Early Church and the Afterlife: Post-Death Existence in Athenagoras, Tertullian, Origen, and the Letter to Rheginos*, 168
Rashi (Rabbi Schlomo Yitzchaki), 41, 43, 149; *The Pentateuch with Rashi's Commentary*, 148, 169
realized eschatology. See eschatology, realized
recollection. See remembrance
recurrence, 93, 119–21, 176–77
redeemer, 60, 81, 84, 87, 91, 129, 166. See also savior
redemption, 3, 10–12, 53, 61–64, 69, 72, 74, 79–82, 85, 90–91, 94, 96, 98, 102–3, 105–6, 156, 166, 172. See also salvation
"Redemption and Resurrection, 4Q521," 166
Refutation of All Heresies (Hippolytus), 27–28, 70–71, 77, 91, 144, 159, 161–62, 165
reincarnation, 108, 112, 173–74
"Reincarnation" (Ullman), 174
Religious Experience and New Materialism: Movement Matters (Rieger and Waggoner, eds.), 135
remembrance, 37, 53, 62–64, 79–81, 83, 91, 98, 107
Republic (Plato), 145, 153, 157
Responsibility to Awe, A (Elson), 20, 75
restoration, 60, 99, 102–4, 112–14, 122, 175. See also apokatastasis
resurrection, 3, 12, 61–62, 92–124, 126, 130, 150, 153, 156, 165–72, 174–75, 177–78
Resurrection and Reincarnation (bar Yochai), 174
Resurrection Debate, The (Silver), 166
Resurrection of the Body in Western Christianity 200–1336 (Bynum), 177
Revelation (Biblical book), 95

Rieger, Joerg: *Religious Experience and New Materialism: Movement Matters*, 135
Riedweg, Christoph: *Pythagoras: His Life, Teaching, and Influence*, 158
Rise of Intellectual Liberty from Thales to Copernicus, The (Holland), 142
ritual, 32, 36, 62, 139–40, 151, 173
"Road to Objects, The" (Harman), 3, 136–37
Robinson, James: *The Nag Hammadi Library* (editor), 152, 154, 159–60, 167
Rojcewicz, Richard: "Everything is Water," 20, 142
Rojtman, Betty: *Black Fire on White Fire: An Essay on Jewish Hermeneutics, from Midrash to Kabbalah*, 57, 155
Romans (Biblical letter), 163, 177
Rosen-Zvi, Ishay: "The Wisdom Tradition in Rabbinic Literature and Mishnah Avot," 10, 138–39
Rowe, Christopher J.: "Myth, History, and Dialectic in Plato's Republic and Timaeus-Critia," 36, 146
Rubenstein, MaryJane, 29: *Entangled Worlds: Religion, Science, and New Materialisms* (editor), 135; *Pantheologies: Gods, Worlds, Monsters*, 143–44; *Worlds Without End: The Many Lives of the Multiverse*, 143
Rukeyser, Muriel: "Poem," 129
rulers, 40, 52–53, 162

Saadia Gaon, 7, 100, 115, 118, 122, 168, 175; "Treatise 7, Concerning the Resurrection of the Dead in This World," 97, 168, 175, 177
Saadia Gaon: His Life and Works (Malter), 168, 177
sacrament, 34, 55, 57, 85, 106
sacrificial theology, 61–62
Sadducees, 93–94
salvation, 10–12, 59, 61–62, 72–73, 80–81, 92, 96, 101,103, 105–7, 138, 168. See also redemption
Samuel (Biblical book), 166
sanctuary, 43, 59
sapience. See wisdom
Sappho: Fragment 16, 125
savior, 12, 60, 62, 72, 77–84, 90, 93, 102–3, 153, 155, 164. See also redeemer
Sayre, Kenneth: *Plato's Late Ontology: A Riddle Resolved*, 65–67, 157–58
scars, 115, 123

scatteredness, 11, 43, 50, 57, 78–79, 94, 102, 106, 110–11, 115, 121–22, 128–29
Schlossberg, Eliezer: "From Periphery to Center: Early Discussion of Resurrection in Medieval Jewish Thought," 168, 175
Schneider, Stanley: "Kotnot Or (Genesis 3:21): Skin, Leather, Light, or Blind?," 154
Schutte, Heinz R.: *Weltseele: Geschichte und Hermeneutik*, 138
Schwartz, Dov: "From Periphery to Center: Early Discussion of Resurrection in Medieval Jewish Thought," 168, 175
Schwartz, Howard: *A Palace of Pearls: The Stories of Rabbi Nachman of Bratslav*, 156, 165
science, 1–2, 8, 16–17, 20–23, 25–26, 29, 143
scintillation, 92, 107, 121, 159
Scott, Alan: *Origen and the Life of the Stars: A History of an Idea*, 138, 172
secret, 4, 46, 48, 64, 67, 109–10
"Searching for Wisdom in and Beyond 4Qinstruction" (Goff), 137
The Secret Book of John: The Gnostic Gospel Annotated and Explained (Davies, editor, translator, and annotator), 49, 52–53, 70, 73, 83, 152–53, 160
Secret of Servitude, The (bar Yochai), 173
Secret Revelation of John, The (King, editor, translator, and annotator), 155
seeds, spiritual, 44, 78–80, 84
Seelenfreund, Morgan: "Kotnot Or (Genesis 3:21): Skin, Leather, Light, or Blind?," 154
Sefer Bahir (also *Sefer ha-Bahir*), 51, 152, 173
Sefer ha-Likkutim, 154
Sefer ha-temunah, 47, 151
Sefer ha-Zohar. See Zohar
Sefer Kuzari (Halevi), 41, 148
sefirot, 41, 47–48, 51, 54, 152
Selected Letters of Friedrich Nietzsche, The, 176
The Seminars, Book II: The Ego in Freud's Theory and in the Techniques of Psychoanalysis, 1954–55 (Lacan), 179
Sentences of Sextus, 138
serpent, 6, 128
Seth (Biblical and apocryphal figure), 77, 139
Sethian, 10–11, 40, 46, 49, 70, 73, 139, 154, 158–60
Sethian Gnosticism and the Platonic Tradition (J. Turner), 139, 158–59
Sha'ar ha-Gemul (Nachmanides), 173
Shakespeare (Coleridge), 23

Shaviro, Steven, *Posthumanities: Universe of Things; On Speculative Realism*, 5, 137
Shaw, Gregory, 33–35; "Theurgy and the Platonist's Luminous Body," 140, 145–46, 173; *Theurgy and the Soul: The Neoplatonism of Iamblichus*, 136
Shenker, Orly: "Fractal Geometry Is Not the Geometry of Nature," 89, 164
Shiur Qomah, 150
Sifrei Devarim, 100–1, 166, 169
Silver, Daniel, 94; "Nachmanides' Commentary on the Book of Job," 173; *The Resurrection Debate*, 166
Simplicius, 66, 158
Sirach, 7, 138–39, 147, 154; *also called* Ecclesiasticus, Ben Sira
skepticism (Academic), 4, 15, 136
Socrates, 4–5, 16, 18, 23, 30–34, 64–68, 82, 146
solitude, 71, 73, 91, 101, 159
"Solomon Ibn Gabirol's Doctrine of Intelligible Matter" (Dillon), 70, 159
Sophia, 71, 76, 78, 88, 155, 160–61. See also Wisdom
Sophist (Plato), 31–32, 145
sophistry, 2, 29–30, 32, 144–45
soteriology, 56. See also redemption; salvation
soul, 6, 21–22, 27, 32, 35, 42, 44–45, 47, 50–51, 76, 81, 84–85, 94, 97–99, 101, 103–5, 107–14, 116, 118–19, 122, 124, 142, 149–50, 166–67, 171–74
spark, 57, 112
speculative realism, 5, 15, 137
Speusippus 66, 70
sphere, 48, 109, 113
"Spiritual Formation in Hellenistic Jewish Wisdom Teaching" (Uusimäki), 138
Spiritual Seed, The: The Church of the "Valentinians" (Thomassen), 160
star. See celestial bodies
Statesman (Plato), 65, 67, 157
Stauros (Cross), 84, 163
Stevens, Wallace: "Peter Quince at the Clavier," 119
stoicism, 10, 20, 27, 93, 108, 119–20, 125–26, 173, 176, 178
Story of Christian Theology, The: Twenty Centuries of Tradition and Reform (Olsen), 140
"Structure of the Transcendent World in the Tripartite Tractate (NHC 1,5), The" (Thomassen), 165

successive incarnation. *See* incarnation, successive
Summa Contra Gentiles (Thomas Aquinas), 167–68, 174–75, 177
Summa Theologiae (Thomas Aquinas), 167–68, 174–75
"Summer Harvest" (Valentinus), 162
Surah al Baqara, 136
Symposium (Plato), 30, 33–34, 40, 65, 145–46, 148, 157, 179

Talmud, 54, 93–94, 97, 99, 122, 149, 152, 166, 171. *See also specific titles*
Tanhuma-Yelammedenu (Midrash) (also *Tanchuma*), 41, 43, 148–50, 173, 177
Targum of Jonathan ben Uzziel. *See Targum Pseudo-Jonathan*
Targum Pseudo-Jonathan, 40, 55, 148, 154, 170
Tatian, 171
teacher, 8–9, 11, 44, 62, 67, 71, 81–82, 85, 102, 126
Teachings of Silvanus, 138
Tertullian, 45, 63, 71, 75–76, 82, 84, 95–97, 99–100, 107, 116, 159, 161, 167–68; "Against Marcion," 169; *Against the Valentinians*, 162–63; *On Modesty*, 162; *On the Resurrection (De Resurrectione Carnis)*, 150, 175; *Prescription Against Heretics*, 163
Testaments of the Twelve Patriarchs, 155
Tethys, 18, 141
Thales, 15–21, 23, 27–30, 141–42, 144, 146
Thales of Miletus: The Beginnings of Western Science and Philosophy (O'Grady), 144
Theaetetus (Plato), 18, 65, 67, 141, 157
Theodoret, 73
Theodorus the Atheist, 29, 144
Theodotus, 44, 75, 81, 84–85, 161, 163. See also *Excerpta ex Theodoto*
Theogony (Hesiod), 17, 27, 29, 141
theology, 1–2, 5–7, 9, 12, 16, 18–20, 22, 25, 30, 36, 41, 43, 58, 63–64, 71, 90, 119, 127, 136, 138–39, 142–43
Theology of Arithmetic, The: Number Symbolism in Platonism and Early Christianity (Kalvesmaki), 69, 158–60, 164
theophany, 33, 35
Theophrastus: *On First Principles: Known as His Metaphysics*, 19, 158
theosis. *See* divinization
Thessalonians (Biblical letter), 166
theurgy, 4, 33, 137, 172
"Theurgy and the Platonist's Luminous Body" (Shaw), 140, 145–46, 173

Theurgy and the Soul: The Neoplatonism of Iamblichus (Shaw), 136
Thomas Aquinas, 21, 30, 97, 113, 115–16, 118, 122; *Summa Contra Gentiles*, 167–68, 174–75, 177; *Summa Theologiae*, 167–68, 174–75
Thomassen, Einar, 90; *The Spiritual Seed: The Church of the "Valentinians,"* 160; "The Structure of the Transcendent World in the Tripartite Tractate (NHC 1,5)," 165
Three Dialogues between Hylas and Philonous (Berkeley), 135
Three Essays on the Theory of Sexuality (Freud), 179
"Three Steles of Seth, The," 77
Thus Spoke Zarathustra (Nietzsche), 176
tikkun, 104, 154, 171, 174. *See also* mending
"Tikkun: A Lurianic Motif in Contemporary Jewish Thought" (Fine), 154, 174
Timaeus (Plato), 145–46, 150, 158
Timothy (Biblical letter), 96–97, 167
Tiqqune ha-Zohar, 51, 152
Toorn, Karel van der: "Why Wisdom Became a Secret: On Wisdom as a Written Genre," 7, 137
Torah, 10, 36, 41–43, 94, 138–39, 155, 168, 174. *See also specific books*
Toth, Christopher, 107
Tractates on the Gospel of John (Augustine), 167
transcendence, 1–2, 8, 21, 30, 56, 70, 165
transfiguration, 97, 101, 106–7, 109, 118, 126–27
transformation, 3, 19, 33–36, 38, 45, 49, 58, 60, 63, 85, 96, 101–2, 106–10, 112, 117, 122–28, 178–79
transmigration. *See* reincarnation
"Treatise 7, Concerning the Resurrection of the Dead in This World" (Saadia Gaon), 97, 168, 175, 177
Treatise on the Resurrection (Valentinian letter), 96, 105–7, 124, 156, 167–68, 171–72. See also *Epistle to Rheginus*
Treatise on Resurrection (Maimonides), 168
Trinity, 11, 41
Tripartite Tractate, 71, 75, 77, 79, 86, 102, 159, 161–65, 170
Trouillard, Jean: *Le Mystagogie de Proclos*, 35, 146
truth, 12, 16–37, 67–68, 79, 81–83, 86–88, 96, 103, 105–7, 112, 123–27, 153, 168
Tur HaAroch, 149
Turner, Denys, "How to be an Atheist," 30, 145

INDEX

Turner, John D.: *Sethian Gnosticism and the Platonic Tradition*, 139, 158–59
twoness. *See* dyad
Tyson, Neil deGrasse, 143

Uexküll, Jacob von, 2
Ullman, Yimiyahu: "Reincarnation," 174
unbounded. *See apeiron*
uncreated, 78–80, 87
Underland (Macfarlane), 43, 107
undivided, 41, 45, 68, 73, 77, 88
unity, 39, 45, 57, 60, 62, 66–67, 69–70, 72–73, 86, 88, 90, 104, 106, 111, 172, 174. *See also* monad; One, The
universe, 19, 27, 52, 93, 103, 109, 113–14, 137, 165–66. *See also* cosmos
unknowable, 4–5, 31, 33–37, 43, 49–51, 57, 72, 75, 85, 126. *See also* ineffability
unlimited, 50–52, 66–68, 70, 72–74, 79, 89, 90. *See also apeiron*
Unlimited, 66–70, 73–74, 79, 88, 90–91, 157–58. *See also* dyad; Indefinite
Unlocking the Zohar (Laitman), 173
Untersuchungen über die Entstehung des vierten Evangeliums (Grill), 160
Untimely Meditations (Nietzsche), 176
unwritten doctrines (of Plato), 66–67, 157–58
"Unwritten Doctrines, The: Plato's Answer to Speusippus" (Pepple), 158
utility, 17, 24–26, 28, 89, 130
Uusimäki, Elisa: "Spiritual Formation in Hellenistic Jewish Wisdom Teaching," 138

Valentinian Exposition, A, 60–63, 71–73, 78–79, 159
"Valentinian Theology" (Brons), 161–64
Valentinianism, 11–12, 40, 44–45, 50, 62–63, 70–73, 80–82, 85, 89, 98, 105–7, 126, 159–64, 167, 171
Valentinus, 11, 64, 79, 87, 90–91, 139, 161–64, 167
Vayikra Rabbah. *See Leviticus Rabbah*
Vibrant Matter: A Political Ecology of Things (Bennett), 143
"Virtuous Assent and Christian Faith: Retrieving Stoic Virtue Theory for Christian Ethics" (Cochran), 178
Visible and the Invisible, The (Merleau-Ponty), 57, 129, 154, 179
Vita Anonymi of Pythagoras, 150
Vital, Rabbi Chaim (also Chayyim), 47–48, 54, 111, 151, 153–54, 173

"Vitalist Stopover on the Way to New Materialism, A" (Bennett), 137

Wade, Erik Kaars, 175
Waggoner, Edward: *Religious Experience and New Materialism: Movement Matters*, 135
Weltseele: Geschichte und Hermeneutik (Schutte), 138
White, Stephen A.: "Milesian Measures: Time, Space, and Matter," 142
Whitehead, Alfred North: *Process and Reality*, 15, 140
Whitmarsh, Tim: *Battling the Gods: Atheism in the Ancient World*, 19, 28–30, 142, 144
whole-part relation, 65–66, 68, 77, 81, 84, 86–90, 126
"Why We Like to Know Useless Stuff" (Cossins), 141
"Why Wisdom Became a Secret: On Wisdom as a Written Genre" (van der Toorn), 7, 137
Will to Power, The (Nietzsche), 176
Williams, Michael A.: "A Life Full of Meaning and Purpose: Demiurgical Myths and Social Implications," 139
wisdom, 4, 7, 10, 12, 15–16, 19–20, 24, 31, 33–34, 36, 42, 45, 53, 59, 62, 65, 67, 82, 90–91, 119, 123, 130–31, 137–39, 143, 146, 153, 155, 170, 177, 179
Wisdom (creative figure), 7–13, 33, 36–37, 41–42, 44–45, 47, 49, 53, 60–63, 70–72, 75–82, 84–85, 87, 90, 97, 102–3, 106, 109, 122–23, 126–31, 137–38, 155, 170
Wisdom (Biblical book), 7, 140
Wisdom of the Egyptian Menander, 138
"Wisdom Tradition in Rabbinic Literature and *Mishnah Avot*, The" (Rosen-Zvi), 10, 138–39
Wise and Foolish Love in the Song of Songs (Andruska), 137
Wolfson, Elliot R., 70, 110; *Language, Eros, Being*, 147, 173; "Mystical Rationalization of the Commandments in *Sefer ha-Rimmon*," 151; *Venturing Beyond: Law and Morality in Kabbalistic Mysticism*, 159
"'Words with an Alien Voice': Gnostics, Scripture, and Canon" (P.C. Miller), 164
Works of Saint Augustine, The, a Translation for the 21st Century: Sermons 230–272B, 163
Worlds Without End: The Many Lives of the Multiverse (Rubenstein), 143
Wright, M.R.: "Presocratic Cosmologies," 18, 141
Writing of the Disaster, The (Blanchot), 87, 147

Xenophanes of Colophon: *Fragments: A Text and Translation With a Commentary*, 30, 142, 145

Yaldobaoth, 155
Yitzchaki, Schlomo, Rabbi. *See* Rashi

Zeno of Citium, 27
Zohar (Matt, translator), 41, 43, 46, 48, 51, 56, 100–1, 104, 110–11, 148–49, 151–52, 154, 159, 169, 171, 173–74
Zohar Chadash, 173
Zostrianos, 73, 160

KARMEN MACKENDRICK is Professor of Philosophy at LeMoyne College. Her books include *Failing Desire, Divine Enticement,* and *Word Made Skin.*

www.ingramcontent.com/pod-product-compliance
Lightning Source LLC
Chambersburg PA
CBHW020108020526
44112CB00033B/1098